Conservation Approaches to Earthen Architecture in Archaeological Contexts

Louise Cooke

BAR International Series 2147
2010

Published in 2016 by
BAR Publishing, Oxford

BAR International Series 2147

Conservation Approaches to Earthen Architecture in Archaeological Contexts

ISBN 978 1 4073 0688 9

BAR Publishing is the trading name of British Archaeological Reports (Oxford) Ltd. British Archaeological Reports was first incorporated in 1974 to publish the BAR Series, International and British. In 1992 Hadrian Books Ltd became part of the BAR group. This volume was originally published by Archaeopress in conjunction with British Archaeological Reports (Oxford) Ltd / Hadrian Books Ltd, the Series principal publisher, in 2010. This present volume is published by BAR Publishing, 2016.

Printed in England

BAR
PUBLISHING

BAR titles are available from:

BAR Publishing
122 Banbury Rd, Oxford, OX2 7BP, UK
EMAIL info@barpublishing.com
PHONE +44 (0)1865 310431
FAX +44 (0)1865 316916
www.barpublishing.com

Contents

PART TWO

List of Tables

List of Charts

List of Sketches

List of Figures

CONTENT OF THE CD - Now available as a download (please see below).

The CD contains supporting data referred to in the text. This is paginated seperatly 1-477.

The Microsoft Access Database documenting approaches to the conservation and management of earthen architecture at a global scale.

Please note that the CD referred to above has now been replaced with a download available at www.barpublishing.com/additional-downloads.html

Abstract

This publication summarises research investigating approaches to the conservation and management of earthen architecture.

There are countless forms of earthen architecture found throughout the world. These include: (1) shaped blocks, 'mudbricks'; (2) rammed earth, 'pisé'; (3) placed earth, 'cob'; and for non-load bearing construction: (4) turf and sod construction, and (5) earth placed onto a supporting frame or armature. A number of these different earth-building techniques also make use of earthen mortars and/or earth plasters or renders. In these different forms earth has been used as a building material for domestic, religious, burial, administrative, palatial and domestic structures for the last ten millennia - the legacy is both monumental and vernacular.

Earthen architecture is a class of material, found worldwide, that shares similar properties, maintenance needs and conservation requirements. The similarities associated with earthen architecture make the comparative study of approaches to the material in contexts of use, maintenance, repair, abandonment, conservation, and restoration, useful as a means to reflect upon and assess approaches to conservation.

This research explores these approaches to earthen architecture around the world, and with particular reference to the study area - Iran, Uzbekistan, and Turkmenistan. The different approaches to conservation and management are critiqued in relation to their practical effectiveness, relationship to conservation theory, values of earthen architecture and sustainability. This research explores these complex values: in this respect this work is reflexive, examining our assumptions, to understand what we think and how we feel about earthen architecture.

This research uses the identification of the materials and techniques used for the conservation and management of earthen architecture as a means to understand, articulate and explore attitudes and approaches to the building material, within the context of wider conservation and heritage theory. The proposed transferable framework for earthen architecture is significant as it suggests a more sustainable approach to the conservation and management of earthen architecture. This aspirational framework is concerned with both the practical issues of 'what we do', and the understanding of 'why we do it' within the context of conservation and heritage theory.

Part 1 examines earthen architecture, its study, use, physical properties and more abstract values. Part 2 examines conservation approaches to earthen architecture in archaeological contexts.

The CD contains appendices of supporting data referred to in the main text.

Please note that the CD referred to above has now been replaced with a download available at www.barpublishing.com/additional-downloads.html

Acknowledgemnets

It would have been impossible to complete this research without the help and support of many different people.

I was involved in the the Ancient Merv Project from 2001 – 2005. This work was supported by central funds from the University College London, alongside funding made available through the World Monument Fund. On site the park director Rejeb Dzaparov, and his dedicated team warmly welcomed me. In 2002 our Merv backfilling team made my time on site a real joy, particularly Dowran Durdiev and Chary Dordolyw and their families answered my many questions, and allowed me to intrude into their family homes. Akmohammed Annaev, Myrat Kurbansakhatov and Sonia Kurbansakhatov translated and gave me additional information about contemporary construction practice in the Merv oasis, and Turkmenistan generally. My enormous thanks to everybody who has been part of the Ancient Merv Project particularly Jackie Keily, Gabrielle Puschnigg, Dave Gilbert, Pierre le Brun, Ann Fuerbach, Faith Vardy, Volodya Zavyalov, Sjoerd van der Linde, Henry Fletcher, Nicky Harriss and Nathalie Peek; Kathy Judelson and Gaigysyz Joraev.

I was given the opportunity to visit and work at Çatalhöyük, Turkey in July 2004, this was supported by the University College London Graduate School Research Project Fund. At Çatalhöyük my thanks must go to Ian Hodder and Shahina Farhid and particularly the UCL conservation team, Liz Pye, Dean Sully, Ina St George, Duygu Cleere, James Hales, Trinidad Ricco, and Mags Felter.

I was awarded a generous grant through the Tessa and Mortimer Wheeler Fund administered through the Society of Antiquaries of London. This enabled me to undertake the field work in Uzbekistan.

A generous grant from the UCL Graduate School and Institute of Archaeological Research Fund enabled my attendance at the 9th International Conference on the Conservation of Earthen Architecture held in Yazd, Iran. Thankyou to all those colleagues who were willing to share their experiences of earthen architecture with me.

The UCL Graduate School Conference Fund also funded my travel to Lehm 2004 in Leipzig in Germany to enable me to present a paper on my research on earthen building materials at Merv.

In the early stages of this research The Getty Conservation Institute, sponsored my participation in the Symposium on Reburial on Archaeological sites in Santa Fe in March 2003.

Many other colleagues and earth building experts shared their knowledge and ideas with me; some are cited in the text others are listed in Appendix 3 & 4. All are thanked enormously; errors and omissios are my fault.

My supervisors Clifford Price, Jane Siddell and particularly Tim Williams guided me through the sometimes very muddy waters of my research. Tim particularly offered his support and enthusiasm at all times, and I will be forever grateful.

My family supported me throughout; Dominic Powlesland encouraged me in every possible way imaginable. Alfreda and William have put up with this final conversion of my thesis into a publication. It is at last done!

For my mother who would have loved to be here to see this finished.

Foreword

Earth is a building material found worldwide. The primary component of earthen architecture is soil to which some modification is carried out to improve its workability and durability. There are numerous forms of earthen architecture found throughout the world. The most common uses of earth for load and non-load bearing construction are:

(1) shaped blocks, 'mudbricks';
(2) rammed earth, 'pisé';
(3) placed earth, 'cob';
(4) turf and sod construction, and
(5) earth placed onto a supporting frame or armature.

A number of these different building techniques also make use of earthen mortars and/or earth plasters or renders. These different forms of earth construction are used for religious, burial, administrative, palatial and domestic structures: the legacy is both monumental and vernacular. The immense variety of earthen architecture reflects the geographical diversity and long history of use of earth as a building material.

In September 2001 I was lucky enough to participate in an archaeology and conservation programme at Merv, Turkmenistan. The stunning ancient Silk Road site invites hyperbole with outstandingly well preserved earthen archaeological deposits and awe-inspiring upstanding earth-built monuments.

In 2001 the project built upon the work carried out by Georgina Herrmann, the excitement and drive of a new project director, Tim Williams, and the changing emphasis of work hailed by the listing of the archaeological park on the UNESCO World Heritage List, and World Monument Fund list of 100 sites in danger. This meant the work of a group of archaeologists was no longer restrained to the excavation, recording and survey of the archaeological features of the park, but was also concerned with the conservation of the archaeology and monuments within the park. The unique and challenging aspects of conservation of the ancient earth structures and preserved earthen archaeology were fascinating.

My interest in earthen architecture emerged from my observation of the it's physical properties, and more abstract values and associations of a material perceived to be 'living' and 'breathing'. These were contrasted with other materials perceived to be 'harder', such as stone or fired brick. On the one hand, earthen architecture is perceived to have positive values and associations, such as its ancientness, local distinctiveness and environmental qualities. On the other hand, earthen architecture has negative values and associations, such as its apparent fragility and association with unsanitary conditions. To me it seemed that the multitude and divergent perceptions of earthen architecture must impact upon approaches to conservation and management - was this why some conservation approaches to ancient earthen structures made them look 'new', while some replaced earthen materials with harder, notionally more long-lasting materials, whilst others simply did nothing, leaving the 'unconservable' material in its eroded and eroding form?

This interest led me undertake my PhD concerned with approaches to the conservation and management of earthen architecture in archaeological contexts. I researched the problems, methods, techniques, and results of divergent approaches to the material, looking primarily at sites in Central Asia (Turkmenistan, Uzbekistan and Iran), and I was given the opportunity to see work in the SouthWest USA and Turkey.

Earth remains the building material of choice in traditional societies across the world, whilst in the developed world practitioners have revived the use of this vernacular and ecologically sound building material and reinvented earth as a material suitable for the expression of architectural modernity. This interplay between the contemporary uses of earth in architecture and the potential that these projects have in challenging views and attitudes towards earthen architecture is significant. As such new uses of earth in architecture have the potential to aid in the retention of earth building traditions around the world, and by retaining building craft skills in contemporary contexts these could be called upon for use in conservation and archaeological contexts.

This volume identifies how the perception of earthen architecture impacts the materials, techniques and approaches to the use, conservation and management of the material in a variety of different contexts. In many instances I come back to my experience at Merv in Turkmenistan to explore the issues in detail, whilst the other sites within the case studies provide comparisons of similar or divergent approaches to the building material. By its very nature this research looks at the past, present and future of earthen architecture and I hope will be of interest to anybody concerned with the archaeology and conservation of a material that remains the building material of choice in traditional societies across the world.

Louise Cooke
lucooke@btinternet.com

PART 1

Chapter 1: Introduction

Research Context

This research investigates attitudes and values associated with earthen architecture in order to understand approaches to its conservation and management within the wider context of conservation and heritage theory.

Earth is a building material found worldwide. The similar properties, maintenance needs and conservation requirements of earthen architecture make the comparative study of approaches to the material in different contexts appropriate as a means of exploring practical conservation issues and the wider concerns of conservation and heritage theory. This research is concerned with the approaches to earthen architecture in contexts of use, maintenance and repair (these 'living' contexts in which individuals and communities inhabit earth structures and are engaged in regular maintenance and repair activities), and contexts of abandonment, conservation, and restoration ('abandoned' contexts in which structures exist in an eroded and eroding form, and are sometimes retained through conservation and restoration activities).

My interest in earthen architecture emerged from my observation of the contrast between its physical properties, and the values and associations of a material perceived to be 'living' and 'breathing' when contrasted with other materials perceived to be 'harder', such as stone or fired brick. On the one hand, earthen architecture is perceived to have positive values and associations, such as its ancientness, local distinctiveness and environmental qualities. On the other it has negative values and associations, such as its backwardness, poverty, and association with dirty, unsanitary conditions. To me it seemed that the multitude and divergent perceptions of earthen architecture must impact upon approaches to conservation and management - was this why some conservation approaches to ancient earthen structures made them look 'new', while some replaced earthen materials with harder, notionally more long-lasting materials, whilst others simply do nothing, leaving the 'unconservable' material in its eroded and eroding form? This work seeks to understand these different approaches to the conservation and management of earthen architecture.

My research explores earthen architecture through the compilation of data at global, regional (Iran, Uzbekistan, and Turkmenistan) and site (Merv, Turkmenistan) levels. The dataset used in this research is collated from published sources, unpublished archive material, and site visits to the study area. The dataset documents approaches, materials and techniques utilised for the conservation of earthen architecture. These different approaches to the conservation and management of earthen architecture have been critiqued in relation to practical effectiveness, relationship to conservation theory, values of earthen architecture and sustainability (*Chapters 7-16*). Using the data my research explores and articulates the attitudes and values associated with earthen architecture, within the wider context of conservation and heritage theory.

In reviewing the literature, and visiting the sites used in this research, I always had an overwhelming sense of confusion and desperation: 'What do we do? It's so difficult, there are no solutions, everything is so expensive, I wish I'd never excavated these mudbrick structures… I wish they would just go away…' This research attempts to resolve this confusion through the review of past and present approaches to the conservation and management of earthen architecture. This enables possible approaches for the future sustainable conservation of earthen architecture to be suggested within the transferable intellectual framework for earthen architecture presented in *Chapter 17*.

The collation and analysis of the data concerning the different materials, techniques and approaches used for the conservation and management of earthen architecture, enables us to challenge the perception of earthen architecture as an 'unconservable' material. The question of what is and what is not meant by the 'conservation' of the values of earthen architecture is addressed within *Chapter 17*. The framework for earthen architecture presented in *Chapter 17* is concerned with both the practical issues of 'what we do', and the understanding of 'why we do it' within the context of conservation and heritage theory.

Within the context of conservation and heritage theory this research also explores the basis of 'difference' observed in approaches to conservation and management around the world. The notion of 'difference' has given rise to value judgements, and these have been seen in relation to approaches to conservation: with good conservation characterised by current conservation theory *vs.* bad conservation characterised by approaches 'outside' current conservation theory (Lowenthal 1985). I would argue that 'difference' has often been observed in relation to differences in conservation approaches to 'incomparable' materials, which have very different physical properties, values and associations, such as the conservation of stone and timber structures (discussed below). As such this work extends the debate concerned with the observation of 'difference' in approaches to conservation by focusing on a single material - earthen architecture - which as a broad group shares similar physical properties (see below). The data collected enable me to conclude that conservation activities are contextually dependent, and 'differences' in approaches result from the complex interplay between conservation and contemporary society.

The research goals and questions

Within the frameworks for conservation theory, earthen architecture and sustainability my research had three main goals:

(1) To develop an understanding of current approaches to the conservation of earthen architecture.

(2) To establish a transferable intellectual framework to assist in the conservation decision-making process for earthen architecture on archaeological sites.

(3) To develop an understanding of 'difference' in approaches seen within conservation and heritage theory.

In order to reach my research goals a number of specific research questions had to be addressed:

- Is current conservation theory applicable to earthen architecture?
- Can a transferable intellectual framework for earthen architecture be established?
- Are approaches to conservation dependent on temporal and spatial contexts?
- Are contexts of use, maintenance and repair, and contexts of abandonment, conservation and restoration comparable? How do these affect approaches to the historic and archaeological fabric?
- Can conservation interventions be assessed within their context as a means to better understand our approaches to the historic and archaeological fabric?

and

- Are differences observed in approaches to conservation based on the comparison of materials with widely different physical properties?
- Are differences observed in approaches to conservation based on the assumption that what is advocated by current conservation theory actually impacts conservation practice?

The theoretical framework

The theoretical framework for this work is drawn from conservation theory, earthen architecture and sustainability.

Current conservation theory

Since the development in the 19th century of anti-restoration arguments (*Chapter 7*), conservation theory has been concerned with the importance placed on the archaeological or historic fabric, stressing the importance of authenticity, age, value, and the visibility and reversibility of interventions. Within this context conservation science has emphasised the application of new materials to meet the aspirations of conservation theory. Throughout the 20th century many heritage bodies and organisations have stressed the international importance of 'global heritage', but some heritage commentators (for example, Lowenthal 1985) have criticised approaches to conservation at a global scale, highlighting 'difference and otherness'. Since the 1990s these notions have begun to be accommodated within conservation theory by stressing issues such as intangible heritage, a wider range of values, participation and poverty reduction. The development of conservation theory is discussed in *Chapter 7*, while the characteristics of current conservation theory form the basis upon which the different approaches, materials and techniques for the conservation of earthen architecture are assessed in *Chapters 8-16.*

This work fits within this broader debate concerning the temporal, spatial and contextual basis of conservation (see Denslagen 1993; Theophile and Gutschow 2003), by showing that the physical properties and requirements of earthen architecture are just one factor that impacts upon the materials, techniques and approaches used for its conservation.

Earthen Architecture

The primary component of earthen architecture is soil to which some modification is carried out to improve its workability and durability. There are numerous forms of earthen architecture found throughout the world. For the purpose of this research the classification of these different forms covers the most common uses of earth for load and non-load bearing construction: (1) shaped blocks, 'mudbricks'; (2) rammed earth, 'pisé'; (3) placed earth, 'cob'; (4) turf and sod construction, and (5) earth placed onto a supporting frame or armature. A number of these different building techniques also make use of earthen mortars and/or earth plasters or renders. These different forms of earth construction are used for religious, burial, administrative, palatial and domestic structures: the legacy is both monumental and vernacular. The broad field of study and types of earthen architecture are discussed in *Chapters 2 & 3,* the use of earthen architecture in 'living contexts' in *Chapter 6,* and the patterns of abandonment, erosion and deterioration in *Chapter 5.*

The immense variety of earthen architecture reflects the geographical diversity and long history of use of earth as a building material. Whilst undertaking my research one question that occurred was whether earthen architecture was comparable globally and thus useful in examining approaches and attitudes to conservation. In answering that question I have always felt, yes: whilst earthen architecture is a broad category of material, it does provide a valid, and extremely useful material upon which approaches and attitudes to conservation can be compared. Earth has been used as a construction material for the last ten millennia and is still used throughout the world today; as a result there are countless occupied earthen buildings, upstanding historic earthen structures, and earthen archaeological sites throughout the world. This means that earthen architecture is one of the few materials in which it is possible to examine the approaches and attitudes to conservation at a global scale: many other materials have a limited global distribution, effected by climate, geography and geology, or limited time spans (other materials used globally, such as concrete and structural steel, have only been 'globally' available through the 19th and 20th century, and approaches to the conservation and retention of modern structures are still being established). Similarly, whilst there *is* a great variety and diversity of forms of earthen construction, as a broad class of material, earthen architectural components generally behave and degrade in a similar pattern, requiring the same care, repair and maintenance. The physical properties and characteristics of earthen architecture influence the longevity of structures and subsequent approaches to the retention of the material in contexts of use, maintenance and repair, and contexts of abandonment, conservation and restoration. This is evident on a global scale and so the study of approaches to the conservation of earthen architecture in different contexts is one that is valid for the examination of approaches and attitudes to conservation.

The framework within which earthen architecture is approached in this research is concerned with identifying and articulating the negative and positive values associated with the material - these values are explored in *Chapter 4*, and form the basis upon which the conservation approaches are assessed within the individual site dossiers and within *Chapters 8-16*.

Sustainability
Between 1983 and 1987 the World Commission on Environment and Development was established, and formulated a strategy document (*Our Common Future* - also known as the Brundtland report), this defined sustainable development as:

"development which meets the needs of the present without compromising the ability of future generations to meet their own needs" (World Commission on Environment and Development 1987, 8)

The report was significant in integrating the notion of sustainability and sustainable development within all spheres of life, and some notion of sustainability has become implicit within conservation theory, through the linkage to sustainable development, and emphasis on locality and participation. Contemporary conservation theory has responded to the notion of sustainability and sustainable development through the incorporation of wider economic and social agendas (Clark 2006, 60).

In researching the development and theory of sustainability this research adopts an approach rooted within environmentalism and environmental politics. This criticises the above often-quoted definition as being so simple and vague as to be inherently weak (see Dresner 2002, 64), and recognises that the term 'sustainability' has often been used without a full knowledge of the values it incorporates. My understanding of sustainability is based on the notion of equality between generations - past, present and future (Dresner 2002, 2): recognising and achieving equality between generations, without impacting future decision-making capacities. As such my understanding of sustainability places a much greater recognition on the past than the Brundtland definition. Within this context sustainability is a contemporary value linked to the environment, which recognises the importance of the past, *and* how current use may pose tensions for the future of the resource. My own definition is:

"development linked to the environment, which recognises the importance of the past, and meets the needs of the present without comprising the ability of future generations to meet their own needs"

As such the notion of sustainability adopted through this research contrasts with that advocated through contemporary conservation theory, by envisaging a complex and holistic relationship between people, the environment, and the material remains of the past, for the present and for the future. Therefore the assessment of sustainability encompasses the economics of the conservation approaches (which is much easier to assess and quantify), the physical and environmental impact of the conservation approaches, alongside those broader, more holistic notions concerned with equality between and within generations. The development of approaches to sustainability, and the adoption of the concept of sustainability within conservation theory is discussed in *Chapter 7*, and forms the basis upon which the conservation approaches are assessed within the site dossiers and within *Chapters 8-16*.

Research methodology

The research adopted a three-tiered research methodology:

- Global
- Regional: Central Asia (Uzbekistan, Turkmenistan) and Iran
- Site specific: Merv, Turkmenistan.

Each of these tiers magnifies the depth of contextual understanding.

Global review
Aim - At a global scale this research investigates approaches to the conservation and management of earthen architecture in a variety of different contexts.

Data Collation & Documentation - Information was collated through an extensive bibliographic search concerned with documenting the available published and unpublished material concerning different approaches to the conservation and management of earthen architecture. Additional material was added from conferences, and more informal discussion with peers active in the field of research.

The bibliographic material collated at a global scale is held in a searchable database, and within word documents. The Microsoft Access© database gives a summary of the publication, whilst separating the country, site, and different conservation approaches, materials and techniques. The queries allow site-specific information to be extracted from the dataset.

The database comprises some 575 bibliographic records, concerning work on 320 historic buildings and/or archaeological sites in 51 countries. The bibliographic entries go up to 2006, the database is included on the CD – this enables searches of the global dataset to be undertaken using author, bibliographic reference, site name and conservation approach as search parameters (*Appendix 1*).

The information from the global review is used to provide the background, context and synthesis for the development of the field of conservation theory discussed in *Chapter 7* and earthen architecture discussed in *Chapter 3*. The detailed analysis and assessment of the global dataset is included in *Chapter 3* (charting the development and trends apparent within the field of research), and *Chapters 8-16* (highlighting the different conservation approaches, materials and techniques used at a global scale).

Regional study
Aim - At a regional scale this research investigates contexts of use, maintenance and repair and contexts of abandonment, conservation and restoration of earthen architecture in Central Asia (Uzbekistan and Turkmenistan) and Iran.

The choice of this regional approach reflects the pragmatic and financial realities of undertaking this PhD. I have been involved at work on the archaeological site of Merv, Turkmenistan, and over several seasons this gave me the opportunity to visit sites, towns and villages in the area surrounding the site and the capital Ashkabad. I was also successful in raising funds through the UCL Graduate School to enable attendance at 9th International Conference on the Study and Conservation of Earthen Architecture, held in Yazd, Iran. During the conference and subsequent tour I was able to visit archaeological sites and a small number of settlements. In addition I was awarded a grant through the Tessa and Mortimer Wheeler Fund administered through the Society of Antiquaries of London, to enable me to undertake the field work in Uzbekistan.

Though pragmatic in its origin, the regional approach is a valid one for the study and investigation of earthen architecture. The region has a long and varied history of using earth as a building material for domestic, defensive, monumental, working and religious buildings, alongside monuments connected with funerary practice. The condition of the surviving structures and sites is affected by the extreme continental climate and natural disasters (with all three countries having zones of seismic activity). During the 20th century traditional systems of earth construction changed in these countries, reflecting broader social, economic and political change (discussed in *Chapter 3 & 5*).

Data Collation & Documentation - The review of sites uses published and unpublished information concerning conservation approaches, alongside extensive written documentation and data collected in the form of plans and photographs from field visits. This is supplemented with informal discussion with individuals on site, creating the site dossiers for each site. The site dossiers collate the information concerning the conservation and management of earthen architecture on the sites visited. The dossiers record basic site information, the different conservation approaches, and my discussion and assessment of these approaches in relation to practical impact, conservation theory, values of earthen architecture and sustainability.

For the study area there are 5 site dossiers for Iran, 6 for Turkmenistan, and 6 for Uzbekistan. In addition, supplementary dossiers document sites visited during the course of the research in Germany (1 dossier), Turkey (2 dossiers), USA (5 dossiers) and UK (2 dossiers): these are less substantial than those from the study area; they do not have the same degree of documentation or annotated plans. Together the dossiers comprise information and illustrative material for 27 sites, and contain over 1400 digital photographs taken during the course of this study.

The dossiers are characterised by their pragmatic nature, trying to document and observe as much as possible within a limited time span. For some sites, where my involvement has been for longer periods of time and/or the period of my research has enabled repeat visits, these reflect the depth of information collected and my understanding of the wider context of the conservation and management interventions.

The site dossiers comprise a substantial dataset (the dossiers are included as *Appendix 6*) and a synthesis of this material forms a large part of the discussion of approaches to earthen architecture in *Chapters 8-16*.

Site scale
Aim – To investigate contexts of use, maintenance and repair, and contexts of abandonment, conservation and restoration at Merv, Turkmenistan.

Merv comprises a series of discrete walled cities that developed on adjacent virgin sites. The cities of Erk Kala (founded in the Achaemenid period and occupied to the Seljuk period) and Gyuar Kala (Seleucid, Parthian, Sasanian and Umayyad), Sultan Kala (Abbasid and Seljuk periods), and Abdullah Khan Kala (Timurid to 19th century) and Bairam Ali Khan Kala (19th century) are within the core area of the Archaeological Park of Ancient Merv. The cities survive in an eroded form, comprising the successive layers of eroded and eroding earthen archaeological material, a massive defensive circuit, a number of upstanding monuments and wealth of excavated materials. With its shifting landscape and complex of cities Merv represents patterns of use, maintenance and repair alongside patterns of abandonment, conservation and restoration. The archaeology of the park testifies to a long, rich and diverse tradition of building with earth in the region, which is still reflected in the contemporary construction practice in the area.

The purpose of the site-specific work has been to utilise the skills, expertise and knowledge I have gained through my involvement in the Ancient Merv Project (from 2001- to present) to gain a more detailed and in-depth knowledge of some of the issues associated with the conservation and management of earthen architecture. This dataset collated from Merv provides a basis to understand in more detail the complexity of approaches to earthen architecture in contexts of use, maintenance and repair, and contexts of abandonment, conservation and restoration.

Data Collation & Documentation - The data collation from Merv is concerned with a number of different aspects that impact upon the approaches to the conservation and management of earthen architecture.

(1) In the first instance information from Merv has been used to create a site dossier, within the same framework as the regional review (see above, and *Appendix 6*). The site dossier records basic site information, information concerning the different conservation approaches, and my discussion and assessment of these approaches in relation to practical impact, current conservation theory, values of earthen architecture and sustainability. This site dossier includes just a subset of the digital photographs I have collated whilst undertaking work on the site. The site dossier allows detailed discussion of the temporal and spatial contexts in which conservation occurs, and how contexts of use, maintenance and repair and contexts of abandonment, conservation and restoration affect approaches to earthen architecture.

In addition to the basic level of information recorded and collated within the site dossier work at Merv has also been vital for other aspects of this research.

(2) The study and identification of the inclusions present within a sample of earthen building materials from different locations across the site.

This work was carried out in order to identify the inclusions present within a small sample of the earthen building materials. An assessment of the effect of these inclusions on the longevity and survival of earthen architecture was also considered. The small-scale sampling of earthen materials was carried out in 2003 and the subsequent analysis in 2004.

These data are used within *Chapter 2* to explore the issues concerned with the nature of the inclusions used within the different forms of earthen architecture, emphasising the importance of identifying the components of earth building materials for both the archaeological understanding and conservation of earthen architecture. The accompanying report is included as *Appendix 2*.

(3) The study and survey of historic photographs to explore factors effecting deterioration and erosion of earthen architecture.

There is extensive photographic coverage for the monuments within the Archaeological Park of Ancient Merv, documenting the condition of a selection of the monuments, from at the earliest 1890, to the present day. A survey of the historic photographs from a selection of monuments, and their replication as a means to create 'point-in-time' records was carried out in 2003 and 2004. The photographs provide a fixed temporal reference to understand the nature of factors affecting the condition of the monuments in contexts of abandonment and conservation in the Merv Oasis. This has enabled the identification, analysis and synthesis of those factors that have effected the survival and deterioration of the selected monuments.

These data are used within *Chapter 5* to assist in identifying the factors that effect the erosion and deterioration of earthen architecture in contexts of abandonment. The report is included as *Appendix 5*.

(4) The study and survey of the contemporary utilisation of earthen architecture to explore values, associations and approaches to the material in 'living contexts'.

Earthen architecture is still used as a building material for contemporary construction in the villages and towns surrounding the archaeological park. Data were collected through undertaking a number of informal questionnaires with park staff and local villagers concerned with patterns of use of different building materials in the Merv Oasis. This was concerned with establishing basic information concerning social, economic and cultural factors affecting the utilisation of different building materials. In addition to these questionnaires I undertook a simple survey in two village locations adjacent to the Merv Archaeological Park. In each instance I was concerned with documenting and recording the current utilisation of the different building materials. This information is used alongside more general background information to better understand the attitudes, values and associations of earthen architecture.

These data have been synthesised and used to explore the utilisation, and values associated with earthen architecture in living contexts within *Chapter 6*.

Research contributions

This work provides examples of the various different approaches, materials, and techniques used for the conservation and management of earthen architecture in a variety of different contexts.

The outcomes of this research are:

- An extensive searchable database of worldwide sites, detailing different conservation interventions and supported by an annotated bibliography.

- Site dossiers (bibliographic information, plans and photographs) designed to support the regional and site-specific studies.
- Terminologies, glossaries and recommendations concerning the conservation and management of earthen architecture in archaeological contexts.

Alongside:

- The study of inclusions within archaeological, historic and modern earthen building materials from Merv.
- The study of historic photographs from Merv, used to explore factors effecting deterioration and erosion of earthen architecture.
- The study of contemporary uses of earthen architecture at Merv.

The original contributions emerge from the discursive synthesis concerned with the assessment and critique of conservation approaches for earthen architecture, developed through the site dossiers, and in *Chapters 8-17*:

- The assessment of earthen architecture in light of current conservation theory.
- The use of earthen architecture to assess notions of 'difference' in approaches to conservation.
- The investigation and assessment of contexts of use, maintenance and repair and contexts of abandonment, conservation and restoration of earthen architecture.
- An overview of the contexts in which the conservation of earthen architecture has occurred.
- Recommendations concerning sustainable approaches for the conservation of earthen architecture through the formation of a transferable intellectual framework for the conservation and management of earthen architecture in archaeological contexts for the 21st century.

This research is a significant contribution to the conservation and management of earthen architecture as it collates much of the earlier research and work concerned with the problems posed by the conservation and management of earthen architecture. This provides future decision makers with the ability to make decisions concerned with the practical issues of 'what we do', and the understanding of 'what we do' within the context of conservation and heritage theory.

Publication structure

The publication is in two volumes.

Volume 1.
Part 1 discusses earthen architecture:

Chapter 2 identifies the different forms and components of earthen architecture and, using material from the study of the inclusions present within earthen building materials from Merv, identifies the great variability of inclusions present within earthen architecture (*Appendix 2*). *Chapter 3* explores the development of research into earthen architecture, encompassing the archaeology and conservation of earthen architecture, and the use of earthen architecture in new construction. *Chapter 4* explores the values and associations of earthen architecture, both in the past and present, and those currently being developed. This chapter is based upon my literature review, fieldwork, and discussions with practitioners and experts (*Appendix 3*).

Chapter 5 explores issues of the abandonment and deterioration of earthen architecture: the first part of the chapter draws on past anthropological and ethno-archaeological research to understand processes of change and abandonment. The second part of the chapter explores factors that result in the erosion and deterioration of earthen architecture, to understand both the physical properties of the material, which can make conservation solutions difficult, and to understand the process of archaeological deposit formation. This chapter draws on data collated through the study of historic photographs of monuments at Merv (*Appendix 5*).

Chapter 6 explores earthen architecture in living contexts, examining aspects such as maintenance, repair and renewal, as well as the transfer of skills associated with earth construction. This chapter utilises data from the global and regional studies, alongside more detailed study of the process of contemporary construction at Merv (*Appendix 4*).

Part 2 is concerned with conservation approaches to earthen architecture.

Chapter 7 discusses how theories that developed in the Western Europe in the 19th century have been adapted and adopted by international heritage bodies and still form the basis of current conservation and heritage theory. This chapter provides the in-depth information concerned with the development of current conservation theory, and highlights the characteristics of current conservation theory that form the basis upon which the conservation approaches are assessed in the site dossiers (*Appendix 6*), and within the synthesis in *Chapters 8-16.*

Chapters 8-17 are concerned with the synthesis and assessment of conservation approaches for earthen architecture. *Chapters 8-16* discusses the conservation approaches adopted for earthen architecture through the global dataset and the regional site dossiers (*Appendix 6*). The different conservation approaches are classified as - backfilling, capping, consolidation, 'do nothing', drainage, encapsulation, maintenance, reconstruction, removal/ relocation, restoration, sheltering, stabilisation, and undercut repairs. These different solutions are assessed in terms of practical effectiveness, relationship to current conservation theory, values of earthen architecture and sustainability. *Chapter 17* returns to my research goals summarising my understanding of current approaches to the conservation of earthen architecture; establishing a transferable intellectual framework for earthen architecture on archaeological sites; and understanding 'difference' in approaches seen within conservation and heritage theory. The chapter concludes by answering my research questions and highlighting areas of potential for future research concerned with earthen architecture, conservation and heritage theory.

Volume 2
The *second volume* on CD contains appendices of supporting data and the Microsoft Access Database documenting approaches to the conservation and management of earthen architecture at a global scale.

Please note that the CD referred to above has now been replaced with a download available at www.barpublishing.com/additional-downloads.html

Chapter 2: Earthen Architecture

Introduction

There are countless forms of earthen architecture found throughout the world. These reflect the very local nature of the building materials and building techniques, and the multitude of ways in which earth can be manipulated and used as a building material. This geographical diversity and long history of use lends an overwhelming diversity to earthen architecture throughout the world.

Earth as a building material

The most common uses of earth for load and non-load bearing construction: (1) shaped blocks, 'mudbricks'; (2) rammed earth, 'pisé'; (3) placed earth, 'cob'; and for non-load bearing construction: (4) turf and sod construction, and (5) earth placed onto a supporting frame or armature. A number of these different earth-building techniques also make use of earthen mortars and/or earth plasters or renders. In these different forms earth is used as a building material for domestic, religious, burial, administrative, palatial and domestic structures: so the legacy is both monumental and vernacular.

The materials used for earthen architecture are dependent on a local geology that gives access to soils suitable for use in construction (see below). Other factors influence the type and nature of construction, for example, access to space for mudbrick manufacture and drying, access to timbers for shuttering and/or timber frames, and length of dry season for allowing materials to dry out. In addition all of the earthen structures may be composites using stone, timber and fired brick as determined by local geology, climate, building style and social, economic and political contexts of construction (Rapoport 1969).

Shaped blocks, 'mudbricks'

Mudbricks use a well-mixed wet earth, which is normally (although not always) combined with a good quantity of vegetable matter, most commonly straw or chaff. The earth mixture is then formed either by hand or cast in wooden moulds to make regular sized bricks. The casting might occur on a bed of straw, or straw may be placed on the top of the mudbrick once it is cast (Horne 1994). The moulded bricks are then stacked in the sun to dry, the period a brick is left to dry is dependent on local climate, and local customs, for example in Iran lines of bricks are left to dry for 4 or 5 days, and then turned on edge to finish drying, for perhaps another week (Fig. 1-8; Horne 1994; Khalili 1983). Once dry masonry techniques are used to construct walls, vaults, arches and domes, utilising earthen mortars, and the surface may or may not be coated in earthen (or other) plasters and renders.

Shaped blocks of earth can also be formed or moulded by hand, resulting in a shape that varies from spherical, cylindrical to ovoid. Hand shaped mudbricks are still preferred in many places as the fingerprint impressions and joint mortar result in a more coherent and stronger wall (Houben and Guillaud 1994). Mudbricks cast in moulds may be of a variety of shapes and sizes dependent on the size of the mould used, local building technique, and construction, and can also vary dependent on the location in the building (Damluji 1992, 128-132). Mechanically cast mudbricks using forms and presses, rapidly produce uniformly shaped and sized (and often more compact) mudbricks.

Evidence for mudbrick manufacture and use is geographically and chronologically diverse. General evidence seems to suggest the earliest forms of earth construction utilised placed or rammed earth techniques, but with mudbricks adopted from the early Neolithic onwards, developing from the earliest hand shaped mudbricks, through to more standardised, and cast or moulded mudbricks (Helwing 2003; Campbell and Pyrce 2003). Early evidence of this technique is found through the fertile crescent of present day Iraq, Syria, Iran and Turkey, with the earliest mudbricks formed by hand, either as hand modelled clay balls (within the Pre-Pottery Neolithic A at Jericho c. 10,200 – 8,800BC); or cigar shaped bricks as at Nemrik and Mlefaat (both Iran) (Helwing 2003). There followed a trend for more standardised, but extremely large bricks (c.1m in length), then moulded/cast mudbricks are found from the late Aceramic Neolithic (c. 8,000-7,600 BC) and final pre-pottery Neolithic (7,600 – 6,900BC) (*op cit*). From the Samarra period onwards cast mudbricks were very standardised, but tended to remain large, reducing in size throughout the 5th millennium BC, until a standard brick size was established in the Uruk period (*op cit*). The standardisation occurred concurrent with the construction of the first monumental structures, such as the Uruk Temple and Walls (Iraq), associated with the Heirachical society (with the decreased mudbrick size linked to the utilisation of forced labour for manufacture and construction of the large monumental complexes (*op cit*)).

Following on from this development mudbrick construction has been found and continues to be practiced throughout the world (see Kemp 2000 for detailed discussion of the development of the technique in Egypt). Mudbrick construction was utilised by the Romans with construction and use described by Vitruvius (Book II. Chapter III), stating the sorts of clay, time most suitable for manufacture and sizes of bricks for different types of structure. With evidence of Roman mudbrick construction found throughout the Roman Empire (Seefried 2004). The technique was then re-introduced from North Africa to Spain in the 8th century, known as *atob*. The Spaniards took their own mudbrick technique to the Americas in the 16th century, adding it to an already rich and diverse earth building culture in South and Central America (Argumendo 1981). In the United Kingdom the clay lump and clay bat buildings of Cambridgeshire, Essex, Norfolk and Suffolk attest to the introduction of variations of the technique from the continent in the 18th century (McCann 1987).

In Iran, Turkmenistan and Uzbekistan mudbricks are utilised for both load and non-load bearing construction.

The use of mudbrick has continued in the study area through to the present day for a majority of structures, only changing in the course of the 20th century in response to social, cultural and economic change.

In Iran mudbricks (*khest*) are used alongside earthen mortars (*gel*), and coated in renders and plasters (*kahgel*) (Beazeley and Harverson 1982; Horne 1994). The first handmade plano-convex mudbricks date from 8450BC, at Tappeh Ozbaki and Ganj Darreh (both Iran), where shaped blocks of mud were put one on top of the other to form walls (Vatandoust 2003). More contemporary practice records that the prepared earth is placed within open sided wooden moulds (of varying sizes depending on local tradition). Once the mould is filled it is carefully removed and the mudbrick left to dry, prior to stacking and use in wall construction (Horne 1994). Earth mortars and earth plasters are worked through the addition of water and other materials to the basic earth mix (see below).

In Turkmenistan and Uzbekistan, box shaped earth blocks (mudbrick) (*gala kerpic*) and hand shaped oval earth bricks (*guvalja*) are used alongside earthen mortars (*palsyk*), and coated in earthen plasters and renders (*suwoq/suwool gelina*). The earliest mudbricks in the area are those associated with the Jeitun culture (c. 6000BC), and are hand shaped oval bricks (c. 20-25cm wide, 60-70cm long) (Reutova and Shirinov 2004). From the second half of the 4th millennium BC rectangular mudbricks begin to appear, and rectangular moulded bricks, used with earth and straw plasters appear at the end of the 3rd millennium BC at Sapalli, Gonur and Togoluk (all in Turkmenistan) (*op cit*). Similar to the developments recorded in Mesopotamia, the standardisation enabled the construction of monumental and defensive structures (*op cit*).

Rammed earth, 'pisé'

The rammed earth technique of earth construction uses a relatively dry (or semi-dry) mix of earth placed into shuttering and rammed till half of the height to achieve a hard, compacted mass. In damp weather the moisture content of the soil will be sufficient, and in other instances a maximum of about 10% water is added to the earthen mix to assist in compaction (Walker *et al* 2005). The soil is shovelled into shuttering, and rammed in lifts. Once the material is dried, the formwork is removed to a new horizontal or vertical location ready for the next layer. As a general rule rammed earth may contain little or no modifying material, other than that which is already present in the earth, as there is generally less shrinkage to counteract (*op cit*). Rammed earth may be constructed with and without surface renders and plasters, dependent on local practice and custom (Fig. 9-11, 31).

General evidence seems to suggest the earliest forms of earth construction utilised placed or rammed earth techniques, but with mudbricks eclipsing these techniques from the 8th to 4th millennium BC (Helwing 2003). An analogous process is noted with earthen material combined with gypsum and crushed pottery formed within planks in deposits from the 4th millennium BC at Uruk (Iraq); and the technique is associated with the Longshan Culture in China from 2500BC (Lui 2004). Evidence of the technique is spread, as with mudbricks, globally and in particular in Yemen, Morocco, the Iberian Peninsula and China. Pliny describes the process in his Natural History:

"Have we not in Africa and in Spain walls of earth known as 'formacean' walls? From the fact that they are moulded rather than built, by enclosing earth within a frame of boards, constructed on earth side. These walls will last for centuries, are proof against rain, wind, and fire, and are superior in solidity to any cement. Even at this day Spain still holds watch-towers that were erected by Hannibal…" (Pliny's Natural History Book XXV, chapter xlviii, cited Williams Ellis 1920).

Later in the 13th Century Moses Maimonides gives a description of earth construction as:

"The builders take two boards, about six cubits long and two cubits high, and place them parallel to each other on their edges, as far apart as the thickness of the wall they wish to build; they steady these boards with pieces of wood fastened with cords. The space between the boards is then filled with earth, which is beaten down firmly with hammers or stampers; this is continued until the wall reaches the requisite height, and the boards are withdrawn." (Moses Maimonides, cited Simpson 1892, 700).

The distribution of rammed earth may be linked to access to timber for the shuttering formwork. In the United Kingdom the rammed earth buildings of East Anglia attest to the influence of 18th century agricultural reformers. Today rammed earth techniques are significant in the 20th century use of earth throughout the world, exploiting the environmental and aesthetic qualities of earth (see Walker *et al* 2005).

Placed earth, 'cob'

Placed earth techniques of earth construction use a moderately wetted earth, which is built up in freestanding lifts without the use of formwork. In this form of construction the earth is moderately wetted, and may be mixed with chaff or straw. This is treaded and kneaded until it is a soft, cohesive, plastic mass. The mixed earth is passed on to the wall builder, thrown up in large balls, or using forks. This mixed earth is then either placed or forcefully thrown directly onto the wall. Wall construction occurs in lifts, (often of a height equal to that between the hand and the elbow), and each lift pared down with a flat backed spade to form a straight face. Each lift is left to dry prior to commencing the next vertical lift. Techniques of construction comprising placed earth are evident both with and without surface renders and plasters, dependent on local practice and custom (for example there is enormous variation in the *chineh* and *paksha* techniques utilised in the study area (see below).

Generally techniques of using placed earth have been poorly identified, and less is known of the origin and spread of this technique, in comparison with mudbrick techniques. As with rammed earth general evidence seems to suggest the earliest forms of earth construction utilised placed or rammed earth techniques, but with mudbricks eclipsing these techniques from the 8th to 4th millennium BC (Helwing 2003). There is the tentative suggestion that a process analogous to *paksha* was used alongside the earliest mudbricks in association with the Jeitun culture in Central Asia (c. 6000BC) (Reutova and Shirinov 2004).

Placed earth is the earth building technique that is most commonly identified in the southwest United Kingdom, (Cornwall, Devon, Somerset, Dorset and Hampshire), although historic buildings in the East Midlands, Solway Plain and documentary evidence for non-load bearing walls on the Yorkshire Wolds attest to a much wider distribution prior to industrialisation (Best 1642; papers collated Hurd and Gourley 2000). The various regional forms in the United Kingdom use a stone plinth for a foundation course and load-bearing and non-load bearing walls rely on a wide thatched roof to cast rainwater away from the wall body (Fig.12-15).

Within the study area there is evidence for both rammed earth and placed earth (*chineh* in Iran and *paksha* in Turkmenistan and Uzbekistan) used for load and non-load bearing wall construction (Hermann (1999) also claims they are used for roof construction). Unfortunately, the discussion of the development of these techniques through the archaeological and historical evidence is complicated by the failure to identify the subtle differences between the two earth building technologies, and tendencies in the literature to classify any earth building technique that is not mudbrick as *pisé de terre* or rammed earth (particularly so when all we see is the evidence of the wall, rather then the process of wall construction) (for examples of this confusion see Wulff (1966) and Herrmann (1999)). This confusion is added to by the local terms in which *chineh* and *paksha* refer universally to any earth wall that does not comprise mudbricks, regardless of the nature of its original construction (conversations in study area; *pers comm.* Horst Schroeder).

In Iran high *chineh* walls are used for both load bearing and non-load bearing construction (Fig. 16, 19, 34-35). Sir John Chardin describes the technique as follows:

"the Wall is built by Layers, which they let to dry, before they lay a new one on, and it is built so, that the higher it rises, the narrower it grows" (Sir John Chardin *Travels in Persia* 1673-1677, 259).

These walls are built up in lifts of quite well sorted earth. In some instances the individual lifts are plastered in an earth plaster, before the erection of the next lift (Walls 2004). Layers of fired brick or mudbrick are sometimes inserted between each lift to produce decorative patterns and assist in erosion resistance.

In Turkmenistan and Uzbekistan earth used for construction of *paksha* walls is broken up, wetted and worked by a team of craftsmen, the wet earthen material is thrown up in clods to the master builder who throws the clod down and shapes the earth into a battered angle for the lift. This is then pared down with a slightly curved-backed spade. The lifts decrease in height up the length of the wall, and a section c.1m is completed and angled diagonally to join the wall section (Fig. 17, 18, 20, 36). Lengths of reeds and/or wood are placed between each of the *paksha* lifts to add seismic resistance to the structure (recorded Bukhara (Uzbekistan) - *Appendix 6)*. Gustav Krist describes the technique as:

"they make great balls of mud, pile them up, and stamp them down with their feet. When one layer has dried out the next is placed on top" (Krist 1937, 87)

This technique is used for the construction of domestic and monumental structures. The vertical and horizontal construction lifts and bands associated with this technique accounts for some of the characteristic erosion patterns visible on earth walls, and also for the misleading description of this technique as one in which 'blocks' of earth are used in construction (Herrmann 1999).

In addition to this placed earth technique, a solitary non-load bearing boundary wall was recorded in Nurata (Uzbekistan), which showed the characteristic patterns of lifts, and marks for the original formwork expected in rammed earth construction (Fig. 9 & 10). The make up of this wall was very different to that seen in the placed earth walls, with the lifts comprising of poorly sorted, random aggregate and dry earth (the wall was again referred to locally as *paksha* – although this time distinguished as '*paksha* made by women' (Conversation Nurata (Uzbekistan) *Appendix 6*).

Turf and sod construction

Turf and sod construction uses cut blocks of topsoil which are given strength by the vegetation root mat. These cut blocks are then stacked root-to-root and grass-to-grass to add cohesion and strength to the wall (Fig. 21-22). The techniques of turf and sod construction are widely spread both historically and geographically. Examples exist in Scotland and Iceland where restoration and rebuilding of historic turf structures is an active interest of heritage bodies and organisations (Walker and McGregor 1996). Agricultural settlers in the 19th century also used turf for construction on the plains of North America. Generally, techniques of turf and sod constructed have been poorly identified and less is known of the origin and spread of this technique. This may be because this class of material has a tendency to form non-monumental archaeological sites. In addition this widespread technique is simple and, as with placed earth techniques, its very abundance and mundanity may have been a factor in its exclusion from critiques of earthen architecture.

Earth placed onto a supporting frame or armature
Earth placed onto a supporting frame or armature uses a moderately wetted earth, mixed with chaff or straw until it is a soft, cohesive, plastic mass. The mixed earth is applied in thin layers (or balls) to a wooden frame or armature, which acts as the load bearing structure.

As with some of the other forms of earth construction, techniques of using earth placed onto a supporting frame or armature have not met with great interest and are generally excluded from critiques of earthen architecture. We can assume the technique is chronologically and geographically widespread. In the United Kingdom techniques of placing earth on a supporting frame (wattle-and-daub and mud-and-stud) are found in both the archaeological and historic building record. Daub panels were the common method of infilling traditional timber frame buildings, used up until the 17th century, for both internal and external walls (Fig. 25, 32).

Within the study area the techniques of using earth as infill on a supporting frame is used in Uzbekistan (*sintch*). Here load-bearing timber frames are infilled with either mudbricks or hand-shaped ovate earth balls (Fig. 27, 39). In other instances fired brick, breezeblock and cement are also used as infill for the *sintch* structures (Fig. 23, 24, 37). The exterior of these structures is then sometimes coated in an earthen plaster and render (and more latterly harder, cement-based render) or otherwise the infill material is left exposed (*Appendix 6*).

Earthen mortar
Earthen mortar is the material used to bond brickwork for those techniques which use shaped blocks of earth and masonry to construct walls, vaults, arches and domes. Earthen mortar may also be combined in a number of other earth building techniques such as placed earth.

Earthen mortar can consist of the same earth that the earth blocks are composed of, such as a well-mixed wet earth, which is then combined with a good quantity of vegetable matter, most commonly straw or chaff. However a slightly different earthen mixture is often used for mortar, with variations of water content or organic matter depending on local tradition. As the mortar is a component part of the process of using shaped blocks of earth, the discussion on the uses, development and spread of earthen mortar has occurred in the same contexts of those appropriate to mudbricks (see above).

Earthen surface finish
Earthen surface finishes (plaster and render) are used on the interior and exterior of earth construction comprising mudbricks and earth placed on a supporting frame or armature, and sometimes (although not always) on rammed earth and placed earth.

Those structures (even mudbrick structures) that are not lived in, such as cattle and storage areas, may not have an exterior or interior plaster (for example Horne 1993). Like earthen mortars, the earthen surface finish can consist of the same earth as the earth blocks. However, a slightly different mixture is often made up for the surface finish, one of higher water content and different organic materials, or at least different quantities of the same organic material (see below). As the surface finish is a component part of using earth as a building material, the discussion on the uses, development and spread of earthen finishes has occurred with reference to the other building forms.

The earthen surface finish acts as both a protective and decorative layer for the earthen building substrate. As the effects of weather erode the protective layer it can be repaired and renewed through maintenance (see *Chapter 6;* Fig. 28). It is also this functional interior and exterior layer of earth plaster that is decorated through painting, carving or incising techniques. Other surface finishes include rubbing and polishing of the interior and exterior of earthen walls, and/or the additional coating of walls with plant or oil-based materials (see Houben and Guillaud 1994).

Within the study area earth plasters are used alongside lime plasters (*gaunch*) for both surface protection and decorative detailing (Fig. 26, 38). Lime plasters applied over an earthen plaster are often perceived as longer lasting than earthen plasters (Horne 1994), and the use of lime plasters is also linked to social and cultural perceptions and assertions of economic power (Damluji 1992; Jerome 2000).

An increasing number of structures have harder, cement-based renders applied to the exterior surface. The use of these materials has been perceived as reducing the need for maintenance (Fig. 33, 54). However, it is well documented that where these cement based renders have been applied they create a harder, impermeable barrier under which there is an increased rate of erosion and deterioration (see *Chapter 9, 10 & 14*).

Corners
Corners on earth structures are generally the weakest points of construction, and those points most at risk from subsequent erosion and deterioration (Chapter 5). As such corners often have higher foundation courses or high plinths (in fired brick or stone). In addition some earth structures (particularly in temperate, wetter climates) utilise stone or brick quoins to anchor adjoining building facades.

The three main forms of earth construction form corners differently. For mudbrick structures corners depend upon the type of brick bond, but often utilise interlocking bricks. For rammed earth, corners pose slightly more problems, requiring the moving of the formwork, or in contemporary earth construction can use special corner formwork sections. Both mudbrick and rammed earth structures often build corners first in order to ensure the correct alignment of the wall. For placed earth construction, corners are formed going around each of the construction lifts, and pared down in line with the plinth at the completion of each lift.

Variety and identification of earth construction methods.

Enormous variation and diversity exist within the classification of different earthen building techniques. These are affected by local custom and climatic and environmental factors, together with function and variation in source materials (Hughes 2002). In some instances many of these techniques can be found in combination. Such is the case with the Icehouses at Merv, where alternating bands of *paksha* lifts are used alongside mudbrick, and the defensive *qala* walls at Shahdad (Iran) in which mudbrick is inserted between the *chineh* lifts, whilst the *sintch* buildings of Uzbekistan have a great variety of materials mudbrick, earth plasters and fired brick used for infill.

Some variations of the different forms of earth construction may be intended to alter the properties of earthen architecture in order to assist and assure the longevity of the structure. For example the benefits of placed and rammed earth are such that erosion will occur within the individual lift and not affect the lift either below or above it (this is at it most successful in the individually plastered layers) (*pers comm.* Archie Walls).

The identification of the different techniques can be complicated on historic buildings and on archaeological sites, due to the very variable and numerous local adaptations of earthen architecture techniques used. Understanding the methods by which walls are constructed can assist in this identification. For example evidence of coursing and mud mortar will assist in the identification of mudbrick walls, rammed earth may be distinguished by the regularity of shape of lifts and the presence of formwork marks, and placed earth may be distinguished through the presence of (or voids left by) organic materials between the lifts, the variable dimensions of each lift, and the presence of shovel marks on the battered exterior faces (see Hughes 2002). Further evidence may be gained through the analysis and identification of the mineral and organic components and inclusions within the earthen substrate (see below).

The type of earth construction used is reliant on geology and geography, whilst the materials and components of earthen architecture vary depending on locality and tradition (see below).

The primary component of earthen architecture is soil. Soils comprise mineral (sand, silt, clay) and organic components, and derive from the decomposition and weathering of parent rock materials (see Limbrey 1975). The role of sand in making a soil suitable for use in earth construction is its ease of use (when compared with clay-rich soils), and role in limiting the amount of shrinkage and cracking on drying (Rosen 1986) (although too much sand weakens the bricks and causes them to crumble (Fathy 1973)). Silts are chemically inactive primary minerals derived from rock (Limbrey 1975). Few soils consist entirely of sand or silt as soil formation implies the presence of secondary minerals (*op cit*). As such the key components of the soil that influence its suitability for use in earth construction are the mineral components, and in particular the type and characteristics of the clays present. Clays are natural aluminosilicates – secondary minerals that result from the mechanical weathering of rock and formed by the interaction of primary minerals and the soil solution (*op cit*). The different arrangements of these mineral layers result in the different clay types (*op cit* 22-27). Kaolinite, illite and montmorillonite are the clay types most associated with the use of earth as a building material. These different clay types have very different characteristics and these determine the suitability for use in earth construction (Houben and Guillaud 1994). For example, kaolinite is generally stable in water, and is a characteristic of mature, highly weathered, heavily leached soils that do not change volume on wetting and drying (Limbrey 1975, 213), whilst illite is not very stable in water and suffers swell, and montmorillonite is much less stable in water, and expands considerably on wetting and shrinking on drying. These swelling clays are a characteristic of immature and unleached soils, such as those deposited through or disturbed as a result of glaciations (*op cit*).

Sub-soils with a clay content between 20-30% are generally found to be the most suited for mudbrick construction, whilst modern building codes require 25-45% clay in mudbrick manufacture (Rosen 1986), and 5-20% clay content for rammed earth (Walker *et al* 2005). A problem with the assessment of the particle size distribution of soil is the great variability imparted by other factors that influence the suitability of soil for use in earth construction, such as maximum particle size, plasticity, shrinkage, organic matter, and soluble salts (for a full list of factors see Houben and Guillaud 1994; Walker *et al* 2005). As a result the different components of earthen building materials are influenced by the characteristics of the soils and clay present, and measures taken to moderate and change the soil characteristics.

Much of the analysis of archaeological and historical earthen building materials indicates great variation between material properties, alongside an overwhelming variety of inclusions within the basic earthen mix (Brown *et al* 1979; Lewis 1980; Boyer 1990; Jerome 1993; Sharma *et al* 1995; Casoli *et al* 2000; Bazara 2004). With modification many soils are suitable for use in construction, in a number of cases the earth available from the most local source is unsuitable for use in construction, and in this case geology and local tradition determine the location from which the earth is quarried. It is also the case that different sources of earth may be utilised for the different building components, for example differences in the earths utilised for different forms of earth construction (such as mudbricks and earth plasters) and different types of construction, such as the sourcing of earths for the manufacture of *tamdyr* ovens in Turkmenistan (*pers comm.* Gaigysyz Juriev).

MODIFIER	ROLE AND FUNCTION	REGION/ COUNTRY USED	REFERENCE
AGGREGATE (BONE / GLASS/ CERAMIC etc)	There are various different purposes for the use of ceramic fragments within earthen building materials. For example: (1) grog and temper make mudbricks stronger (2) potsherds inserted into the mortar as a key for subsequent plasters; (3) pottery is inserted into the face of walls to protect from failing water (4) potsherds used in vaulting to fill interstitial spaces during construction and to bond the bricks together without mortar to reduce the risk of collapse and failure associated with shrinkage; and (5) symbolic or ritual purposes. These may also be included within the earthen mix for their associative values - aggregates reused within the Chapel of Reconciliation (Appendix 6).	Generic	McIntosh 1974, 1977; Fathy 1973; Markovic 1993; Searle 1912; Davey 1961; Kemp 2000.
ANIMAL or HUMAN HAIR	As a binder to keep cracked lumps of soil in place.	Generic	Hughes 1983, 179.
ANIMAL DUNG	Reduces the plasticity of the soil	Generic	Hughes 1983, 179.
ASH	Makes the clay component less susceptible to shrinkage and swelling.	Generic	Hughes 1988, 1043
ASPHALT/ BITUMEN/ PITCH	Makes water resistant and adds strength by reducing absorption and evaporation. (Used in the White Temple at Uruk c. 3500BC; Ibn Battutah describes pitch being used as a building material in Baghdad.)	Mesopotamia Generic	Walker and McGregor 1996b; Campbell and Pyrce 2003
BLOOD	Waterproofing agent	Generic	Hughes 1983, 1988
CARBONATES	Added to harden earthen materials through a pozzolanic reaction to increase soil strength and reduce moisture damage Carbonates may be picked up from ashy occupation material and/or soil with high carbonate content chosen for the earth mix.	Worldwide (eg Near East/ American Southwest).	Rosen 1986; Matero 1999; Walker and McGregor 1996b
CALCITE and CALCAREOUS SOILS	The crystallisation of calcite on drying provides a framework within which shrinking and swelling can be accommodated (Limbrey 1975, 213). The effect of weathering of calcite is to breakdown the clay platelets so they are re-aligned to create a more impenetrable mass	Generic; Iran, Baghestan	Horne 1994, 130; Thomas 1999; Espinosa 1993.
CHAFF	Chaff is added to the basic earthen mix in order to reduce shrinkage and swelling of the clay component.	Generic	Horne 1994, 137
DUNG	Added initially as a plasticiser and subsequently as a binder inhibiting the dispersion of the clay in contact with water. Another reason for using dung may also be that the plant material is already considerably broken down, in contrast to the use of raw plant material which requires breaking down into smaller components.	Generic	Ashurst and Ashurst 1988
FERMENTED MILK and STRAW	Produce residual cellulose chains that act as reinforcement agents and the fermentation products (polysaccharides) produce bonding effects in the soil.	Generic	Hughes 1983.
HAEMATITE	As a surface polish which will act as a waterproofing agent.	Generic	Hughes 1983.
LENTILS	Stone ground yellow and/or black lentils added to Khorasan (mortar). Acts as a cohesive mix when wet and smoothed out to a durable and marble-like glossy finish.	Generic	n/a
LIME	(see carbonates)	Generic	n/a
MANIOC JUICE	Reduces attack by ants and termites as a poisonous coating.	Generic	Hughes 1983.
PALM OIL	Waterproofing agent	Generic	Hughes 1983.
PEBBLES	Similar to temper – aggregates in the form of pebbles creates a 'skeleton' for the fine-grained particles to stick to and limit the amount of shrinkage and cracking when the mudbrick dries.	Generic	Rosen 1986.
RICE FLOUR	Used to impart strength and hardness, on account of the starch content (starch is insoluble in cold water).	Generic	n/a
SALT	There is an association between salt and the prevention of insect infestation, for example Pliny recommended soaking an unfired brick in salted water to prevent weevil attack. Its use is also documented in different locations around the world to stop moisture rise and prevent insect infestation.	Yemen, Iran etc	Damluji 1992
SEAWEED		Cyprus and Scotland	Thomas 1999
SHALE		Tepe Nush-I Jan (in the lower courses).	Lewis 1980

SHELLS	Act as aggregate – may also be calcite rich.	Peru	Gil 1997
SHELLAC	Resinous secretion used for polishing, resin binder.	Generic	n/a
STARCH	Insoluble in cold water, derived from corn, wheat, potatoes, rice and other cereals.	Generic	n/a
STRAW	Straw (normally wheat or barley straw) is added to the basic earthen mix either 'fresh', after a period of rotting or in the form of manure. The importance of the time-lag may be associated with increased ease of use of the softer, wetter material and may also be associated with its fermentation products adding beneficial properties to the clay component of the basic earthen mix (Hughes 1983). Observations from the Merv samples and from fieldwork indicate the great variety of quantities included and the size of the straw included (varying from very fine (less than 0.5cm) to much larger, coarser inclusions between c.5cm-8cm). Various different functions are attributed to the use of straw within earthen building materials, these include: • Counteracting the effects of shrinkage and cracking by holding the material together (Hughes 1983, 178). • As a binder that reduces the soil plasticity (*op cit*). • Helps soil dry out (*op cit*). • Provides bonding when repairing (*op cit*). • The fibrous materials improve the tensile strength of the final product (Torraca 1981, 101). • Current use of straw at Merv is attributed to increasing the ease of working with the earthen mix – the use of straw means that is does not stick to the tools used! • Physically binds and chemically strengthens the clay by adding humic acids (Rosen 1986, 76). • Strengthens against tensile forces The benefit of the uses of straw is summarised as: "*It is known that clay bricks need straw as a binding agent or to be stabilized with sand – at least 30 percent; without this they will crack. The straw fibres seem to hold the brick together while it is shrinking during the process of drying. In the case of mud plaster made with straw, it would be interesting to see whether its observed water repellent properties are due to a simple binding effect, or whether by some chemical change such as the formation of lactic acid during fermentation or whether the water-repellent property of the straw itself, some of which is exposed on the surface of the plaster. It has been noticed that after rain the clay surface of such plaster is washed away and the straw is left exposed over much of the surface*" (Fathy 1973, 224).	Generic	Rosen 1986 Fathy 1973 Hughes 1983 Davey 1961 Horne 1994
WOOD POWDER	Added to mud mixture to increase tensile strength and resistance.	Iran	Vatandoust and Mohktari 2004.

Table 1. Inclusions within architecture.

Inclusions within earthen architecture

In almost every earth-building tradition some modification and change to the basic earthen material is carried out to improve the workability and longevity, and to limit the volume changes caused by the expansion and contraction of the clay component on wetting and drying (Hughes 1988). There are a great variety of materials added to the basic earthen mix, although their benefits are generally poorly understood (see table 2).

The materials added to the earthen mix are determined by the local geology, availability of materials, custom, and tradition, alongside type of construction, for example more straw may be added to mudbricks which are to be used in dome and vault construction (see Fathy 1973, 9), whilst placed earth (*chineh* and *paksha*) construction in Iran, Central Asia, and in particular Merv, contains little or no straw, in contrast to placed earth construction in the United Kingdom (*cob*) that relies on the inclusion of barley straw.

In a number of instances the presence of inclusions may be accidental rather than functional, and is affected by factors such as the source and quarry of the basic earth used, with recycled materials such as pottery present in earthen architecture formed through the recycling of earlier building materials. In addition factors such as the speed of manufacture may influence the presence or absence of accidental inclusions - with those earthen building materials manufactured more rapidly and more easily with the quarrying and re-use of earlier eroded and eroding building materials, or with rapid manufacture not screening and removing inclusions containing a greater number of accidental inclusions within the earthen mix (see below).

In other instances inclusions may be deliberate and symbolic rather than functional. The incorporation and re-incorporation of material and inclusions within new construction can be interpreted as highly symbolic, with the new construction being associated and linked with a structure in the past (with the incorporation of former building materials and/or incorporation of ancient eroded and eroding building materials) or other symbolic function. For example, in a contemporary context the Chapel of Reconciliation in Berlin (Germany) is concerned with the act of remembrance and incorporates material from earlier construction within the modern rammed earth walls (*Appendix 6*). In other instances the importance associated with the re-use and incorporation of inclusions within earthen building materials may be more associated with religious or ritual functions. For example evidence from 18th Century India suggests the symbolic and associative meanings of the re-use and recycling of building materials:

"I herewith enclose you some of the plaister [sic] I picked up, which had fallen from Hyder's tomb stone. It is said to be composed with earth from Mecca, or as it is called, the Scrapings of the Dust from the Holy Tomb of the Prophet, and consequently must possess many rare and invaluable virtues." (James Kirkpatrick letter to this father, cited Dalrymple 2003, 77).

This suggests that inclusions serve a symbolic function linking a structure with people or places in the past. There is additional evidence from living contexts and from the patterns of deterioration of earthen architecture that indicate that quarrying and re-use of building materials for practical and associative purposes are significant factors effecting the construction, maintenance and longevity of earthen architecture (*Chapter 6*).

Merv

A number of these inclusions could be classified as deliberate (as they have obvious benefits, such as straw), whilst others could be accidental (as they add no obvious benefits, such as glass working waste). These accidental inclusions may be associated with the different sources quarried for the basic earth material.

The identification of the inclusions within the earthen building materials from Merv shows the very variable nature within single structures, across the entire site and between different earthen building types and techniques.
The identification of inclusions within the earthen architecture at Merv indicates a relationship between the type of earthen building technique and the inclusions within them. For example *paksha* generally contains ill-sorted and slightly larger inclusions such as pot and ceramic building material (CBM) fragments, whilst having no plant material. In contrast, mud mortar and mud plaster generally contain well-sorted plant materials and well-sorted smallish pot and CBM fragments. The inclusions within the mudbricks are much more variable and these reflect the much higher number of mudbricks sampled alongside the variable contexts of construction.

The variable nature of the mudbrick inclusions probably indicates where material was quarried and used from, alongside the original investment and care taken in the original manufacture. For example the Parthian mudbricks generally contain least inclusions, whilst all others are likely to contain a variety of different inclusions. Seljuk bricks are more likely to contain a variety of different inclusions, and also more likely to not contain straw. The great variety seen within the Seljuk mudbricks does not seem to vary depending on the nature and function of the structure, although there is a general pattern for a greater variety of inclusions (including the absence of straw) within the mudbricks sampled from the defences and from the palace structure in Shahriyar Ark.

This is perhaps indicative of the sources and quarries used for mudbrick production, (perhaps influenced by rubbish disposal within the Seljuk city (*pers comm.* Tim Williams)), alongside the nature of the mass mudbrick production and labour organisation (Rosen 1986). The fact that this pattern is seen in the Seljuk mudbricks for the defences and for the high status palace structure is surprising, and contrasts with previous studies which see a relationship between the quality of mudbricks (as measured through homogeneity and make-up) as indicative of the status of the building (for example Rosen 1986).

Chapter 3: Study of earthen architecture

Archaeology, conservation and earthen architecture.

The awakening of interest in archaeology in the 19th century led to the beginnings of both the excavation and conservation of earthen architecture. In 1887, in present-day Iraq, Robert Koldeway developed techniques of tracing to enable the identification and subsequent archaeological excavation of mudbrick structures (Matthews 2003). This represented a significant shift from a mining approach to archaeological discovery. These techniques of archaeological identification and excavation were further defined and refined in the course of the 20th century in Middle Eastern archaeology (*op cit;* Kenyon 1981). These developments, alongside work in other locations (for example, McIntosh 1977), in related archaeological disciplines (for example, Rosen 1986) and through ethnoarchaeological work (for example, Horne 1994), have created current identification, excavation and interpretation techniques and methodologies for excavated earthen architecture (for example, Hughes 2002).

The developing interest and technological grasp of the archaeology of earthen architecture occurs at the same time as interest in the conservation of examples of upstanding earthen architecture is developing. Cosmos Mindeleff carried out the first conservation work on the Great House at Casa Grande Ruins, Colorado (USA) in 1892 (Matero 1999; Matero *et al* 2000). This work consisted of the repair of the eroded wall bases using fired bricks set in a cement mortar, and in 1903 the first redwood and corrugated iron roof shelter was erected over the structure with $2000 provided by the United States Congress (*op cit*). The use of shelters and the methods of repair of undercut wall bases developed at this site are still used for the conservation of earthen architecture in both historical and archaeological contexts worldwide.

The study of the conservation of earthen architecture developed and diverged at an international scale in the last half of the 20th century. In 1966 ICCROM established a scientific programme for the conservation of earthen architecture, and approaches were developed through a partnership between the University Museum, Philadelphia and the Italian Archaeological Institute in Baghdad and Turin (Carter and Pagliero 1966). Preliminary testing of a variety of chemical consolidants for the treatment of wall surfaces was undertaken (such as sodium silicate, calcium chloride and polyurethane resins) as well as the manufacture of new stabilised mudbricks in Italy and Iraq (Torraca *et al* 1972). At the conclusion of this work, those involved summarised, "These tests merely showed that the problem was a large one and that far more than a few scattered experiments were needed before a serious study could be initiated with any hope of success" (*op cit* 260).

Research from the 1960s onwards is best represented in the available literature. This work is primarily concerned with materials analysis, the effects of moisture and the use of chemicals for both the modification of new materials and the surface treatment of archaeological or historic fabric. The work carried out in this first experimental phase characterises the type and nature of international collaboration and research (Balderrama 2001), and is documented in synthetic volumes (Houben and Guillaud 1994; Warren 1993, 1999), and in the proceedings of international conferences concerned with the study and conservation of earthen architecture. These represent the bulk of the literature concerned with the conservation of earthen architecture (see below Chart 1-4; Matero and Cancino 2002). These published approaches to earthen architecture are influenced by both funding and perceived international priorities by the organising committees. These conferences therefore do not reflect the majority of the work carried out concerned with the study, conservation and management of earthen architecture, but do reflect aspects of current practice and interest. Since 1972 the Terra conference series has represented the bulk of the literature concerned with the conservation of earthen architecture (Matero and Cancino 2002). The 1st (1972) and 2nd (1976) International Conferences on the Conservation of Mud-Brick Monuments were held in Yazd (Iran). At this stage general concerns were conservation approaches on archaeological sites and structures alongside the use of traditional and chemical consolidants. By the 3rd (1980) held in Ankara (Turkey) a number of themes were highlighted. These included the long term efficiency of chemical protection methods, the use and design of shelters, reburial, laboratory testing, adapting historic structures, alongside 'institutional' aspects, such as the need for sponsorship and the collation of bibliographic information (Alva *et al* 1980).

The 4th meeting held in Lima (Peru) in 1984, with the 5th in 1987 held in Rome (Italy), emphasised the use of traditional and chemical treatments, issues related to decay, and issues related to the demise of vernacular architecture. The ideas raised within the context of this meeting were significant as they represented the emergence of the debate on the efficacy, suitability and use of 'traditional' approaches as possible alternatives to chemical treatments for the conservation of earthen architecture (Nardi 1987, 77).

More recently the 6th (Las Cruces, New Mexico, 1990), 7th (Silves, Portugal 1993,) 8th (Torquay, UK 2000), and 9th (Yazd, Iran 2003) conferences have mushroomed in their size and representation. 'Solid' themes have now been established that these conferences return to. These include the history and traditions of earthen architecture, conservation and restoration of buildings and archaeological sites, seismic mitigation, applied aspects (such as research concerned with chemical properties), the retention and renewal of earth building techniques, alongside the political, legal and economic contexts of conservation.

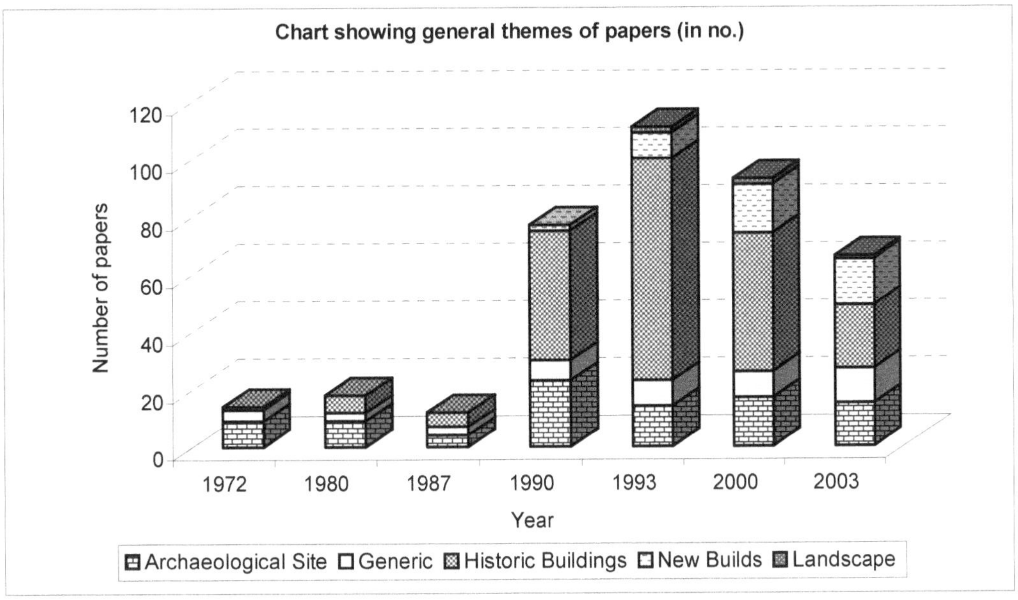

Chart 1. Chart showing general themes of the papers presented at the international conferences concerned with earthen architecture (by number).

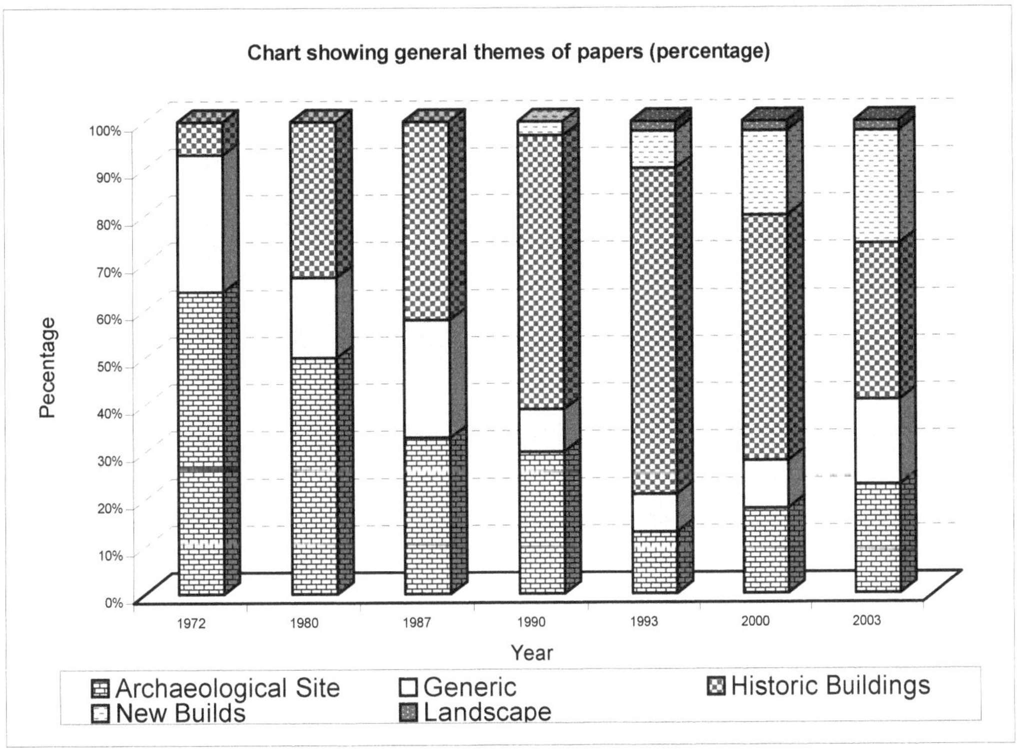

Chart 2. Chart showing general themes of the papers presented at the international conferences concerned with earthen architecture (by percentage).

The charts show the great increase in the frequency, size and coverage of the international conferences from 1990 onwards – archaeological sites have always been significantly represented but are eclipsed from 1993. From 1990 onwards studies concerned with historic buildings were most represented, similarly from 1990 onwards, new earth construction, and landscape studies were included within the broad topics covered by the presented papers. n.b 1976 and 1984 omitted from table.

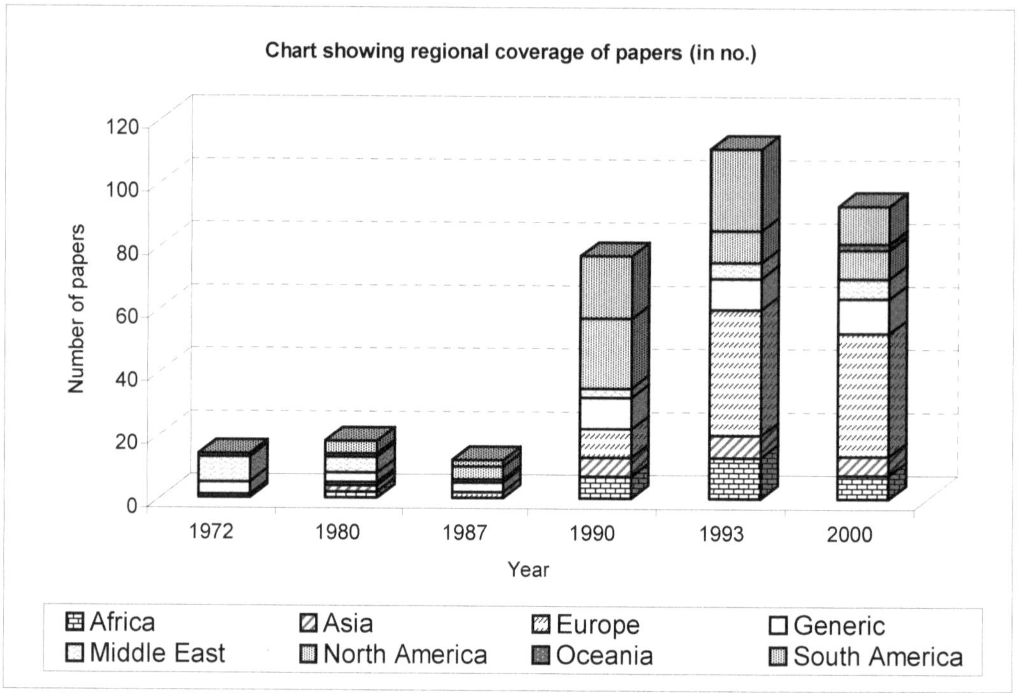

Chart 3. Chart showing general regional coverage of the papers presented at the international conferences concerned with earthen architecture (by number).

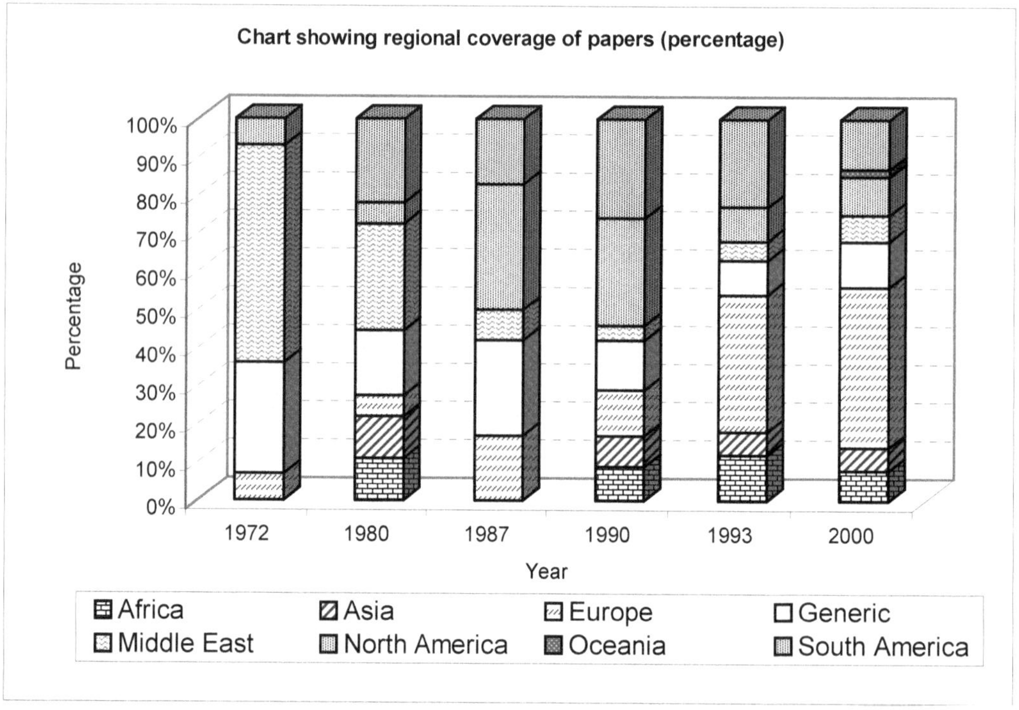

Chart 4. Chart showing general regional coverage of the papers presented at the international conferences concerned with earthen architecture (by percentage).

The charts show that earlier conferences were generally more concerned with work carried out in the Middle East, and with the number of papers presented concerning working in South America, Europe and North America increasing though time – with work in the Middle East almost completely eclipsed from 1987 onwards. Work in Asia and Africa is generally poorly represented in comparison with work in Europe and the Americas. The generally small number of 'generic' papers indicates that these conferences generally cover active, project-based research rather than broader issues concerned with the conservation of earthen architecture. n.b 1976 and 1984 omitted from table.

A great variety of different materials and approaches have been used at different periods, different sites, different monuments and different phases of the same monument. Today the study and conservation of earthen architecture remains characterised by very diverse approaches, although following general patterns of change in current conservation theory these approaches are rooted more firmly in the involvement of local communities, collection of local knowledge, participation and planning of conservation activities as part of wider site management plans (following similar broad trends in conservation and management, see Matero *et al* 1998, *and Chapter 7*).

Use of earth in new construction

The archaeological study and conservation of earthen architecture has borrowed from, and been influenced by the use of earth in new construction (and vice versa). The skills and techniques associated with earthen architecture were 'grey arts' which were handed down from generation-to-generation in both formal and informal apprenticeship contexts. In response to the changes from the agricultural and industrial revolutions the traditional basis of some of these skills has changed (see Brunskill 1981 for a discussion of the decline and revival of vernacular styles and materials in the United Kingdom that can be seen as a model for elsewhere).

As with many fields of study a more formal interest and 'revival' of earthen architecture developed from the 18th century to the present day. For example, Georg-Claude Goiffen described the rammed earth method of construction in *L'art du macon pisuer* (1772) (McCann 1983), and Francois Cointeraux took up the construction method as a modern and rational approach to building in *Ecole d'Architecture Rurale* (1790) (*op cit*). Cointeraux was driven by enlightened ideals to provide low-cost, rural housing by using and exporting rammed earth techniques for general construction, and in locations in which earthen building techniques different to rammed earth were practiced, or in which earth-building traditions had declined. Cointeraux emphasised the Roman origins of *pise de terre* in France, and this stress on the classical origins of the building material was important in the technique gaining social acceptance (McCann 1987).

These ideas were translated and copied, for example by David Gilly in his *Handbuch der Land-Bau-Kunst* (1811), and by Henry Holland in his *Communications to the Board of Agriculture* (1797) (McCann 1983). This first spurt of 'formal' interest in earthen architecture was concerned with agricultural reforms, and driven by needs for social improvement and environmental consciousness. The effects were comparatively minor, when set against the backdrop of rapid agricultural change, and industrialisation that occurred in the 18th and 19th century. However a number of structures were built, including cottages at Woburn Park, Bedford (UK) (*op cit*); less numerous but more high profile structures for aristocratic Europeans, such as the Priorat Palace, Gatchina (Russia) by Nicolay Lvov (Munteanu *et al* 2003); and the utilisation of rammed earth techniques was particularly significant in the new world and European colonies (Williams Ellis 1920).

From the end of the 19th and over the course of the 20th century these interests in earthen architecture have grown and diversified. Worldwide there are many modern architects who have gained inspiration from vernacular earthen building traditions and styles (Oliver 2003). For example, the arts and crafts movement in the United Kingdom was concerned with vernacular forms and techniques, in this context Ernest Gimson, constructed an Art Nouveau cob building at Coxen, East Budleigh, Devon (UK) (1910) (Egeland 1988). At the same time the archaeological discovery and recording of historic pueblo settlements gave the inspiration for the regional architectural style (although not materials) in New Mexico (USA) (Wilson 1997) (*Appendix 6*). Similarly in the early part of the 20th century both Le Corbusier (1887-1965) and Frank Lloyd Wright (1867-1959) designed (although did not execute) earth buildings (Easton 1996), (although Frank Lloyd Wright's California Houses did use the on-site earth materials for concrete block production).

Interest in earthen construction as an available and affordable alternative to other building materials has fluctuated in the 20th century. In the United Kingdom a publicised interest in earthen building materials is seen at the turn of the 20th century, and again at the end of both world wars, with a concern for affordable housing. At the turn of the 20th century St Loe Strachey was at the forefront of a revival and interest in rammed earth as a cheap and locally available building material, his interests manifested in one-off individual projects, for example, experimental structures were associated with the Garden City Movement, 1905, competition for cheap cottages (held at Letchworth), and temporary structures erected at his home (converted to a hospital) during the First World War. These were documented in his magazine, *The Spectator* and by his son-in-law Clough Williams Ellis in 1920 and 1947. These interests influenced the decision of the government to investigate rammed earth as a building material in experimental cottages built in Amesbury, Wiltshire (UK) between 1919 and 1921. The experimental cottages represent a government response to post-war shortages of building materials, alongside the requirements of the post-war land settlement programme (Jaggard 1921; *Appendix 6*). It is significant that in this context the rammed earth cottages were valued as utilising a 'new' material, which if used like concrete, was suitable for the expression of modernity (Swenarton 2003).

The experience of rammed earth cottage construction at Amesbury was that these were much more expensive than the cottages constructed from brick and concrete. This, alongside changed social-cultural perceptions in which traditional buildings and materials did not provide 'homes fit for heroes' returning from the First World War (Burman 1999), influenced the decision at a government level not to adopt rammed earth as a building material.

The adoption of earth building technologies can also be seen worldwide in response to economic crisis throughout the 20th century. For example, with rammed earth walling adopted in the USA in the Great Depression (Patty and Minium 1945; Easton 1996); and the use of earth for construction for private and public buildings within the former East Germany in response to shortages of building materials - (even to the extent that they hoped to export knowledge of earth building to developing countries to earn foreign income) (Schroeder 1993; Rath 2004). In Australia GF Middleton was researching earth construction for the Commonwealth experimental building station (Middleton 1951), and influenced the ongoing adoption of earth and modified earth construction methods in Australia. In the 1970s interest in earthen buildings revived in the context of the energy crisis, burgeoning counter-culture, and environmental back-to-land movement. This is typified by the 1973 Shelter publication (Kahn 1973), and the founding of CRATerre and the Centre for Alternative Technology in the 1970s. The growth of interest, and the revival of earthen building techniques in specific locations worldwide have also been driven by individual architects such as Hassan Fathy in Egypt (Fathy 1973 and Steele 1997); and Nader Khalili in Iran and USA (Khalili 1986). These individual projects and interests occurred to a backdrop of changed architectural traditions for monumental and domestic structures in the post-colonial Islamic world, in which modernity was expressed through concrete and steel structures in an international Islamic Style (see Frishman and Khan 1994 242-272). Both Fathy and Khalili were concerned that this type of construction was inappropriate and a loss to their respective country's cultural diversity.

In the 21st century people are drawn to the environmental, sustainable and aesthetic qualities of earthen building materials. There exist today a number of earth building organisations lobbying for greater awareness and utilisation of the building material (for example *Dachverbandlehm*, see Steingass 2003), with rammed earth in particular utilised as a sustainable material able to perform and function within contemporary architecture (Walker *et al* 2005). With these interests in earth as a building material its use has acquired new values and meanings, for example Martin Rauche utilises the aesthetic qualities of rammed earth as art (Kapfinger and Rauche 2001).

In other respects the study of earthen architecture, has been concerned with the documentation and exposition of the aesthetic qualities of examples of vernacular uses of the materials from around the world (Rudofsky 1964; Dethier 1982; Bourgeois and Pelos 1996). This appreciation of the aesthetic values of earthen architecture was largely in reaction against architectural modernism and broadly categorised as based on the voluminous, continuous, gentle, soft and organic proportions of well-maintained earthen architecture (although both modernism and aesthetic values were combined in Peter Aldington's Haddenham (UK) Cottage gardens – see Brown and Bryant 1999). This aesthetic quality of earthen architecture is now exploited by advertising campaigns, the film, and tourism industry (for example the World Heritage site of Aït Benhaddou (Morocco) is frequently used as a film set).

The development of interest in earthen architecture must be seen within a much wider context of a worldwide decline in the utilisation of the materials and techniques associated with earthen architecture. UN Habitat still estimate 30% of the world's population live in earth buildings, but the nature and type of settlement and land use has altered as a result of complex cultural, socio-economic and environmental changes through the 19th and 20th centuries. The modern construction industry that serves the other 70% of the world's population is characterised by the utilisation of standardised, prefabricated elements, structural steel, fired brick and cement, and has emerged as a key economic power and political lobby. It is within this context of an overall worldwide decline and change in traditional construction, that the interests and study of earthen architecture must be placed.

Chapter 4: Valuing materials and architecture

"When we see a house mental associations rise up in our mind and we judge the house according to our preconceived idea of what a house is. To most people the word 'house' is associated with the word 'home' and to that word exists a whole chain of associations such as 'lasting', 'unchanging', 'real', 'family', 'place to return to', 'solid', 'durable', etc. It has taken probably thousands of years or more of tradition to produce these associations in our minds." (Ronald Duncan 1947, 47).

In his description of building a new rammed earth cottage Ronald Duncan points out that we assign values and associations to structures according to the contexts within which we operate. This is not to say that the values and associations we assign are universal, and indeed the concept and significance of 'home' is one that is both very personal and highly dependent on an individuals context, dependent on both the environmental needs of 'home' and the demands of daily family life (Oliver 2003, 16). The intrinsic value and concepts associated with 'home' are therefore variable.

The associations that we generate are the response to individual experience alongside the norms and expectations of culture and society. It could therefore be argued that our perception of the intrinsic values and associations of building materials is explicitly linked to our aspirations for, and concepts of the 'home'.

Architectural philosophy develops the discussion of the profound effect buildings have on us. De Botton, for example, builds upon Ruskin's arguments about the eloquence of architecture, stating:

"buildings are not simply visual objects without any connection to concepts which we can analyse and then evaluate. Buildings *speak* – and on topics that are readily understandable." (De Botton 2006, 71).

De Bottton argues that buildings talk to us about the life that unfolds within and around them, and how and what we feel about a building is concerned with the values promoted by a building and the lifestyle a building suggests (2006, 72). He goes on to argue that the application of ethics to architecture would enable a better understanding, and creation of, an idealised environment: "In both casual and erudite registers, we are drawn to identifying vices and virtues" (2006, 174).

Many of the values and associations ascribed to earthen building materials and architecture have been developed by comparing and contrasting earthen architecture with other building materials, such as stone or fired brick. As such a value-based approach to discussing the extent to which earthen building materials depart from or match the 'ideal' of a building material is a valid and original approach. In many instances the discussion of the negative (the vices) and positive (the virtues) of earthen architecture is one of counterbalance between different perceptions of the intrinsic worth of the building material.

Earthen architecture: the vices

The comments that followed William Simpson's address to the Society of Arts in 1892 indicate many of the perceptions of earthen architecture that are still present today. Mr Stannus, for example, pointed out that earthen architecture can lack aesthetic value, and be more liable to damage from moisture:

"Mr H. Stannus said when the subject appeared on the paper, many might not exactly see the connection between mud architecture and applied art, but, after listening to the paper, they would be of the opinion that mud, as used in the dwelling of man, was applied art. All arts had arisen from the three necessities of man, viz., food, clothing and shelter; and mud architecture had been exceedingly useful to the shelter of man, protecting him from heat of the sun, from cold, and from fire, though he was afraid, not from damp." (cited Simpson 1892, - discussion following lecture).

The negative view of earthen architecture is one that sees the material as lacking modernity, associated with poverty, backward and uncivilised, cheap, weak, more liable to destruction, linked to ill health and disease, a last resort, and one with unsuitable terminologies.

Lacks modernity

Contemporary building materials and practice are often associated with modernity. This stands in contrast to traditional and local building traditions which are often associated with undeveloped society. This is manifested by the rejection of earthen architecture as a suitable building material: in the developed world through the great rebuilding associated with agricultural improvement and industrialisation (see *Chapter 3*; Walker and McGregor 1996b, 3), and in the developing world where traditional buildings and materials are held in contempt (Oliver 2003, 250). In addition, in the developing world traditional buildings are perceived as substandard, hindering progress and development:

"Largely, the issue is one of prejudice; ignorance and hostility to what have been regarded as 'bush' or 'backward' cultures, antipathy to vernacular architecture and the use of traditional resources and techniques, and fears of being 'held back' from modernizing." (Oliver 2003, 252).

The modern building industry is characterised by the use of "modern materials": structural steel, concrete and fired brick. The projection of these images of modernity, through globalisation, has impacted upon the developing world and influenced government-sponsored and international projects: with a commitment to high technologies for prestigious buildings (*op cit*). As such the perception of modernity has manifested in a rejection of traditional and indigenous forms of construction.

An interesting example of the perception of earthen architecture as lacking modernity comes from New Gourna (Egypt). Hassan Fathy's experimental village is important in defining a new approach to community building, and one that saw an explicit connection between the community and the utilisation of mudbrick for construction. Unfortunately the failure of the project (due to bureaucratic reasons and an apparent lack of community support) provided a setback to the innovative use of earth building materials:

"Because Gourna was never finished, the whole theory of mud brick construction and the attitude to rural housing implied by the use of nonindustrial materials and traditional skills was condemned as cranky and impracticable." (Fathy 1973, 149).

It remains deeply problematic that those involved with building in earth are perceived as outside the norms of contemporary construction practise using 'unconventional' materials.

Associated with poverty
The image of the 'mud hut' has been, and continues to be, seen as the symbolic reference and metaphor for poverty: "Mothers give birth on the dirt floors of mud-brick huts" (Clayton 2005). Even in the recent arguments for fair trade and poverty alleviation at the 2005 G8 summit the iconic image associated with poverty alleviation was one of a solar panel on an African mud hut, increasingly seen as the symbolic reference for the clash of cultures and civilisation:

"And other modern technologies are leapfrogging into the developing world. In some ways, the 21st century arrived before anyone noticed that most of the 20th century innovations never made it! Solar panels sprout on the thatched roofs of mud huts in Kenya." (http://www.channel4.com/science/microsites/M/makepovertyhistory/barefoot/ accessed 8/8/2007).

These ideas are beginning to be challenged within the broader post-colonial context, with the equation between poverty and the image of the mud-hut seen as a simplistic throw back to an era of colonialism:

"not all Africans live in mud huts without electricity or running water … but, so far as the rest of my class was concerned, Africa was a land of mud huts and cannibals. And who could blame them for their beliefs when they'd grown up on Tarzan movies and television series about white adventurers in the bush like Cowboy in Africa and Daktari?" (Eshun 2005)

Backward and uncivilised
Within the political and colonial contexts of the discovery and exploration of other lands, 'other worldliness' and notions of being uncivilised were often associated with the types and forms of indigenous architecture (for example Crinson 1996, 37-71). Earthen architecture represented one of the most significant forms of indigenous architecture, and western society generated perceptions of earthen architecture as backward, debased, uncivilised and nonindustrial.

There was an implicit connection made between industrialisation, development and 'improved' building materials, and between colonialism and the importing of those 'improved' building materials to alter traditional patterns of life (Said 1993; Crinson 1996). This tension between improving and self sufficiency is eloquently described by Hassan Fathy describing Egypt in the 1940s:

"the peasant had been wisely and quietly exploiting the obvious building material, while we, with our modern school-learned ideas, never dreamed of using such a ludicrous substance as mud for so serious a creation as a house." (Fathy 1973, 4).

Even today a tendency still exists for the developed world to perceive and redefine sufficiency (not being dependent on the outside world for subsistence or construction) as underdevelopment (Bourgeois and Pelos 1996, 162). In this respect development programmes and organisations continue to be defined by offering interventions (using the alleged technological superiority of western knowledge) to underdeveloped communities (Oliver 2003; Stohr 2006).

Cheap and easy, rather than good, buildings
The perception exists that earthen architecture produces cheap and easy buildings, rather than good and durable ones. Note, for example, the tone of surprise when commenting cob could be a "surprisingly durable form of building," (Batsford and Fry 1938, 50). This is because earth was perceived as a 'soft' building material, which in comparison with stone or fired brick structures produced substandard buildings.

"By temperate climate standards mud is disconcertingly, even alarmingly soft, a substance so fragile that, when hearing of its use as a building material, many people scoff. How, after all, can one build serious structures presumably no more rugged than that symbol of the ephemeral, a child's sandcastle on the beach?" (Bourgeois and Pelos 1996, 35).

The notional lack of durability of earth architecture is additionally influenced by the relative invisibility of earthen architecture in temperate climates, where many earth buildings are covered over by weather boarding or plaster (Williams-Ellis 1920, 8).

Inherently weak
It is assumed that earthen architecture creates buildings that are inherently weak and have only a limited lifespan. The notion of earthen architecture as more liable to destruction was summed up by Vitruvius, commenting that whilst wattle and daub offered a quick method of construction, it is weak and more liable to damage through combustion:

"As for "wattle and daub" I could wish that it had never been invented. The more it saves in time and gains in space, the greater and the more general is the disaster that it may cause; for it is made to catch fire, like torches. It seems better, therefore, to spend on walls of burnt brick, and be at expense, than to save with "wattle and daub," and be in danger." (Vitruvius Book II. Chapter VIII: methods of building walls. 20)

Vitruvius defined the three conditions architecture must accomplish as: *Utilitas* (utility), *Firmitas* (durability, permanence, resistance) and *Venustas* (beauty) (Vitruvius book 1, chapter 3 section 2). As such the perceived 'vices' of earthen architecture (soft, weak, lacking durability, permanence and resistance) have been defined in contrast to these implicit 'virtues' of a building material (hard, durable, permanent and resistant).

More liable to destruction

Earthen architecture is perceived to be more likely to suffer from rapid destruction:

"They are houses of clay, whose foundation is in the dust, between morning and evening they are destroyed, they perish for ever without any regarding it" (Macaulay 1953, 151).

This is particularly so as in response to natural disasters such as earthquakes, people (and in particular the media) see a connection between earthen architecture, creating sub-standard structures, and structures that are more liable to seismic damage. For example, after the 2003 earthquake in Bam (Iran) questions were raised regarding of the suitability of earth for construction in seismic regions, with blame apportioned to the traditional earthen buildings for the great loss of human life in the earthquake (Branigan and Whitaker 2003; Site Dossier *Appendix 6*). The notion that earthen architecture was easily destroyed was assisted by the dramatic 'before and after' photographs of the Arg-e Bam (http://news.bbc.co.uk/1/hi/magazine/in_pictures/3422997.stm).

As a result difficulties exist in garnering governmental and institutional support to undertake post-disaster reconstruction utilising earthen building materials, and problems are posed undertaking the conservation of earthen architecture in seismic regions (discussion Berlin 2003; Leipzig 2004). However in light of the apparent devastation wrought by the Bam earthquake further research elucidated the type of damage that occurred. This showed that the greatest damage occurred to the reconstructed elements, where the 'new' conservation work had fallen away from the historic fabric (see site dossier, *Appendix 6*). In addition the ability of earthen architecture to withstand seismic damage was highlighted when further work showed that it was the poorest areas of the city, alongside the ruined parts of the Arg-e Bam (but where the earthen architecture had the greatest integrity) that survived better (*pers comm.* Dino Bumbari; discussion Leipzig 2004).

Linked to ill health and disease

Earthen architecture is linked to ill health and disease. It is true that earth structures can pose some specific health problems, such as respiratory diseases associated with falling dust from earth roofs (*pers. comm.* Richard Hughes). However, there is a much more broadly perceived association between ill-health and traditional construction materials. There are also perceptions that traditional earth structures are difficult to adapt to modern sanitary needs (discussion Yazd 2003).

In some instances there is an explicit connection between some diseases and traditional forms of construction and settlement. In Latin America, for example, Chagas disease is associated with rural housing, comprising thatched roofs, mudbrick and mudplaster walls (Bastien 1998). Although relatively unknown, this disease infects 18 million people annually, debilitating and killing adults in the prime of their life. The parasite that causes the disease travels to humans through an infected bug (triatomine) that lives in thatch and wall cracks. Methods to control the disease have focussed on improved housing and hygiene, replacing thatched roofs with tiles, building on concrete platforms, lime plastering, and spraying the interior of structures with slow-release insecticide paints (*op cit* 120). This has assumed a connection between ill health and traditional earthen architecture, in some respects discouraging research into the potential for a link between the illness and other environmental and ecological factors, such as local environmental change and degradation affecting the natural habitat of the triatomine (*op cit*).

Last resort

Earthen architecture has been, and continues to be perceived as a second choice (last option) building material: a "bastard" form of construction (Williams-Ellis 1920, 2). Earthen architecture is perceived as used only as a 'last resort', when geological, environmental, climatic or economic conditions do not allow for construction utilising any other preferable building materials. Often deterministic relationships are envisaged between people, place and geology, as a key factor in the use of earth in construction. Note, for example, the exasperated tone with which Sir John Chardin explains: "The *Persian* Houses are not built of Stone, not because Stone is scarce, but because it is not a proper Material to build with in hot Countries." (Chardin 1673, 257).

Unconservable

Earthen architecture has often been classified as an 'unconservable' material. For example, one of the standard publications on the conservation of historic buildings describes earth as a "despised" material (Fielden 1994, 73). This perception of earthen architecture is linked to the patterns of erosion and deterioration associated with earthen architecture, especially when seen in contrast to other building materials, such as stone. For example, unlike stone buildings which can generally be left in an eroded and eroding form and still leave a visible trace of their existence (albeit so long as the building materials are not robbed), earth structures will, if left in an eroded and eroded form, leave less visible traces of their existence (albeit it may retain archaeological evidence of their existence).

"There is nothing elegant about the temples and fortified monasteries of adobe brick that strew the mountains and deserts of Chinese Turkestan, often crumbling in ruin, often buried in desert sands' (Macaulay 1954, 393).

The notion of earthen architecture as 'unconservable' is also inextricably linked with the development and interpretation of the notion of conservation (see Chapter 7). The notion of 'conservation', and conservation theory, developed primarily in response to the problems posed by the retention of structures comprising fired brick or stone elements, building materials that behave very differently to earth, as such the problems posed by earthen architecture (such as the impossibility of retaining it in an 'as found' condition) create significant tensions when planning for and undertaking conservation.

Unsuitable terminologies
The English language is full of metaphors that underpin negative associations of earthen architecture. The terminology for earthen architecture is often the terminology of 'mud': mud huts, mudbricks, mud buildings. The colloquial uses of the word 'mud' are associated with something worthless or contemptible. For example, the Oxford English Dictionary figurative and extended uses of the word 'mud' give the word as something base or worthless, the dregs or a fool. Phrases that use the word 'mud' in the eighteenth and nineteenth century are concerned with (1) disparaging or slanderous associations (to sling, fling, or throw mud; to drag through the mud; mudflies; mud sticks); or (2) describing something that is unintelligible (as clear as mud). I would argue these negative perceptions underpin the wider negative associations of earthen architecture.

In addition, it has been suggested that there is a symbolic association between stone and the dead, whilst materials such as timber are associated with the living (Bradley 2002, 89). This archaeological theory has historical and modern parallels, for example, shown in the replacement of Lutyen's temporary timber and plaster cenotaph in stone (Ridley 2003, 288-9). A perception exists that stone will last forever, whilst other 'living' materials (in this instance earthen architecture) will decay. Some religious texts make these connections explicit: for example, whilst man is made from mud or clay, he honours his gods as 'living stones' - "like living stones, let yourself be built into a spiritual house..." (1 Peter 2:5). This association between 'living' and 'dead' materials, impacts the appropriateness of the use of different materials for different types of structure (for example, the contrast between domestic and monumental structures). Indeed it is the "living" quality of earthen architecture that we will return to in discussing the one of positive view of the building material below.

Earthen architecture: The virtues

Within the last few decades research has focused upon the social and environmental benefits of earthen architecture (see *Chapter 3*). For some this has fundamentally changed the perception of the value of the material. As with the values associated with earthen architecture in the past, these values are to some extent the result of comparison between idealised norms: in the past, the comparison of 'soft' and 'hard' materials; now between materials perceived to be environmentally and socially acceptable, and those that pose environmental and social problems.

The positive view of earthen architecture values the material as adaptable, aesthetically rich, ancient, autonomous, healthy, locally distinctive, resistant to environmental disaster, linked to humanity, modern, environmental friendly and responsive, and associated with a rich symbolism.

Adaptability
Earthen architecture can be adapted to fit local needs; and structures and monuments can be easily adapted and changed. As materials and techniques are locally sourced, this lends itself to maintenance, repair, renovation and building design changes, and additional rooms or additional storeys can be easily accommodated (Horne 1994; Bourgeois and Pelos 1996; discussion Bath University 2005). In addition earthen architecture can be the material of choice when positioning new structures, as earth buildings become part of the local environment and landscape (Fathy 1973).

Aesthetic
The shapes, colours and texture of earthen architecture create an aesthetic quality to the building material that is unlike others (Kapfinger and Rauch 2001; Walker *et al* 2005). The chameleon-like colour of earth, as it changes from its grey-white hue at dawn, to its harsh yellow in the midday sun, to the burnt orange at sunset, enables earth to respond to, and be part of, its environment (discussion and observations at Merv 2004). The texture of earthen materials also changes: from the smoothness of plastered surfaces, to the rough texture of visible construction lifts. Our response to texture, shape and irregularities generates a positive psychological impact (Weismann and Bryce 2006). These aesthetic qualities are exactly those explored through the use of earth in contemporary sculpture (see below). Similarly, when exposed in plan or section through archaeological excavation, the variation in texture and colour of earthen archaeological deposits show the process of formation, use and deformation.

Ancientness, durability and universality
Earthen architecture has been used for a very long time, evolving and developing at a time in which human civilisation developed and emerged. The archaeological evidence of earthen architecture shows an enormous diversity of forms, and that it can and does last for a very long time (Fathy 1973; discussion Çatalhöyük 2004). In this respect the ambience of earth walls reflects

"durability" and "a feeling of permanence" (Easton 1996, *xi*). This very antiquity confirms the durability and longevity of the skills associated with earth construction. It is a material that exhibits and displays our shared past, and the interconnectivity between people (discussion Merv 2004).

Autonomous

The use of readily available earth as a construction material encourages and allows owner-built construction (Weismann and Bryce 2006). Materials for construction are easily accessible, construction is quick and relatively easy (discussion Bath 2005). Whilst anybody can participate in earth construction it relies on craft skills and techniques, apprentices and master craftsmen, placing value on these crafts skills and intangible heritage at a time and in places where these skills are negated and in decline (Fathy 1973; Houben and Guillaud 1994; discussion Merv 2005). In the twenty-first century the autonomous nature of earthen architecture, enabling construction to occur separate from the global construction industry and markets is a powerful attribute.

Environmentally responsive

The thermo-dynamic properties of earth mean that it is an appropriate building material, which responds to annual and diurnal temperature fluctuations. Hassan Fathy explored these thermo-dynamic properties when choosing mudbrick construction at the new village of Gourna (Egypt). Mudbrick is a poor conductor of heat, but retains absorbed heat for a long period of time (Fathy 1973, 45-47). These thermo-dynamic properties can therefore moderate temperature fluctuations, making earth the building material of choice in desert climates.

More recently increased interest (and legislation) concerned with the thermal properties of buildings has raised interest in the use of earthen materials to moderate the interior climate and enable more comfortable and healthy habitation, which is not reliant on the consumption of fossil fuels for heating or cooling (Weismann and Bryce 2006; discussion Leipzig 2004). Examples of new buildings that choose earthen (in both cases mudbrick) materials explicitly for the positive thermo-dynamic qualities include the Gando Primary School (Burkino Faso) and DRUK White Lotus School (Ladakh, India) (Architecture for Humanity 2006).

Environmentally friendly

Earthen architecture is perceived as an environmentally friendly building material. Environmental considerations in planning new buildings are concerned with the utilisation of healthy, sustainable, energy conscious materials (Dachverband Lehm e.V 2004). As such earthen is perceived to be ideal building material (*op cit*). The environmental qualities of earthen building materials are related to the fact that the material is often quarried and used locally and so transport costs are kept to a minimum. In addition the use and production of earthen building materials minimises the use of fossil fuels and so has a limited contribution to global CO^2 emissions, this stands in great contrast to other building materials such as concrete (discussion Bath University; Pearson 2001; Walker *et al* 2005; Weismann and Bryce 2006). In addition the environmental responsiveness of the building material places less demands on energy use.

Healthy

Modern forms of earth construction in Western Europe are increasingly seen as implicitly connected with good health, through the avoidance of asthma triggers and reduction of respiratory disease through the natural regulation of temperature and moisture fluctuations, and avoidance of the use of chemical synthetic materials within construction (Hawemann 2004; Walker *et al* 2005; Weismann and Bryce 2006). In addition a growing body of research indicates earthen building materials may be particularly efficient at shielding against highfrequency electromagnetic radiation (Dachverband Lehm e.V 2004).

Resistance to natural disaster

The properties of earthen architecture actually make the material more resistant to damage associated with natural disaster. One such example is fire, where the materials are not as combustible as many modern building materials. In addition the benefits of utilising earthen architecture is such that it can limit the emission of harmful chemicals during combustion, this is particularly important as most fatalities in fires are caused by the inhalation of these harmful chemicals (*pers. comm.* K. Harrison). The growing recognition of the benefits of earthen architecture in assisting with limiting the loss of life in the event of fire is shown by the use of earthen building elements in the renovation of the Hôtel d'Orange in Stavelot, Belgium (Thönnes 2004).

Current research on earthen architecture used in modern construction is concerned with adding seismic resistance to earthen building materials (as an important counterpoint to the negative view above). This wealth of research indicates how the incorporation of design expertise and the innovative use of both traditional and modern materials (such cane or geotextile grids) makes earthen architecture a suitable building material in seismic areas (Blondet and Garcia 2003).

Humanity

Earthen architecture embraces, rather than excludes people and community. People and community are needed for quarrying, working, construction, maintenance and adaptation of earthen architecture (Bourgeois and Pelos 1996; discussion Merv). In addition the maintenance of earthen architecture is, in many instances, a community activity often linked to social context (Oliver 2003). Such an example would be the replastering of the exterior of the mosques in Mali (Bourgeois and Pelos 1996).

A connection is seen between earthen building materials, as natural, living and 'breathing', and the people who create and live in them (Dachverband Lehm e.V 2004). As we have already seen symbolic associations between 'living' and 'dead' materials impact the perceived appropriateness of the material.

One of the more philosophical tensions concerned with earthen architecture is the fact in the past the 'living' qualities of the building material were a 'vice' and negatively impacted the view of earthen architecture, whilst today it is those 'living' properties that are a 'virtue' and contribute to the positive view of the material.

Local Distinctiveness
Whilst earthen construction is universal the various construction techniques, materials and forms are rooted in locality, with a local distinctiveness that is extremely diverse. Vernacular architecture (of which earth construction is just one facet) is a significant and unique response to environmental, social and cultural needs (Oliver 2003; Weismann and Bryce 2006). The local distinctiveness of vernacular buildings and building materials, as expressed by earthen architecture, is witness to the diversity and variety of communities around the word and as a key signifier of local distinctiveness. Indeed within the mid-twentieth century earth building revival, for example, the retention of local building materials was urged in order to protect local distinctiveness (Williams-Ellis 1920, 8), whilst similar arguments continue to be voiced today (Weismann and Bryce 2006).

Recyclability
Earthen architecture is a material well suited to reuse and adaptation. The re-use of eroded, collapsed or derelict earth structures in new earth construction enables resources to be recycled (Dachverband Lehm e.V 2004). The reasons this occurs are pragmatic: re-incorporating already worked earthen materials reduces the labour involved in quarrying and mixing, and tidies up the site, clearing away the old to make way for the new. The value of earth as a material that can be re-used and/or returned to its former state makes this a material with only a limited environmental impact.

More symbolically, however, the fact that earth can be re-incorporated in this way allows the values associated with older structures to be incorporated with the new. Within the *Kappelle der Versöhnung* in Berlin (Germany), rammed earth is used as a means to reuse material rather than reconstruct the destroyed former church (*Appendix 6*). For this structure, concerned with the dual acts of worship, commemoration and memory, the choice of earth as the material in construction is profound: it avoids the use of concrete, with its painful association with the Berlin wall, and extends and utilises the values and perception of the material as:

"a natural, living construction material and consequently more easily damaged. The fragility of the structure testifies to the vulnerability of peace and reconciliation. Clay also signifies "healing earth" on the wounds of the location, which should not be "sealed." Accordingly, the liveliness of the material also symbolises the possibility for transformation and the triumph over the location's tragic history." (Braun 2003, 37).

Modernity
With its combination of environmental responsiveness and environmental friendliness earthen building materials are used for the expression of architectural modernity. This allows complex and dynamically beautiful earth walls and structures to reflect modern values of environmental concern (expressed through the eco-friendly building materials), whilst allowing modern construction forms to take place. This is especially the case with modern rammed earth construction (Walker *et al* 2005). Examples of architectural modernity expressed through earthen architecture are within the *Kappelle der Versöhnung* in Berlin, the rammed earth walls at the Eden projects and the beautiful earth walls by Martin Rauche (*op cit;* Kapfinger and Rauche 2001).

Symbolism
Earth is consistently used for symbolic reference in literature, art (e.g. Trotter 2002); music; and language: *nostalgie de la boue*. In addition earth and clay are frequently used for more enigmatic, environmental artistic expression. The influential artist Richard Long uses mud in many of works, he describes the appeal of mud as:

"it's a simple, direct natural material, like water or stones or dust. It's the product of the continual flow of water over millennia, caused by the pull of the lunar tides." (Long 2007, 51).

Long further describes the material as:

"My materials are elemental: stone, water, mud, days, nights, rivers, sunrises. And our bodies are elemental: we are animals, we make marks, we leave traces, we leave footprints" (Long 2007, 53).

In some respects this use of earth as an artists material mirrors the values and associations of earth used as a building material that are listed above, such as its association with humanity and locality. Within the same environmental art tradition Andy Goldsworthy uses mud, dung and clay. Within his enigmatic clay rooms, (enormous and labour intensive installations combining locally dug clay and tons of human hair) perfectly plastered walls erupt when the clay dries and deep cracking forms (Goldsworthy 2000; and see Groom 2003). These works occupy walls or rooms and blur the distinction between the building and the earth (Murray 2007, 10). For Goldsworthy this is significant as these installations explore the relationship between the building and the art work: "these works should feel as if they have risen to a building's surface as a memory of its origin, a connection between the building and its material source (Goldsworthy 2000, 8).

In this respect the use of earth as an artists material can offer a "sensory experience" (Renfrew 2003, 24) which is quite unlike that offered by other materials. Goldsworthy and Long elicit profound responses to the materials they work with and the type of response they generate are significant both as examples of the feelings generated by earthen materials, and *vice versa,* the artistic reflections have a significant impact on our perception of earthen building materials.

'Something else'
In some respects a number of the feelings generated by, and the perception of earth do not fall into discrete categories. In this context earthen building materials offer 'something else', an uncertain something that draws people to it. What makes earthen building materials hold these intrinsic values is uncertain. Perhaps it is the almost universal importance of earth and clay within creation myths and religious texts. Many different cultures from around the world have different variations of creation myths in which clay (or earth) is a motif from which man and woman are formed (such as within the Quran). This 'something else' may also be associated with the fact that today, those people unfamiliar with the use of earth as a building material are consistently surprised and amazed that buildings and archaeological sites have been and continue to be constructed from earth, and earth alone.

Discussion

We assign values and associations to structures according to the contexts within which we operate. As such the values we assign to earthen architecture are both very personal and highly dependent on individual context. The negative (vices) and positive values (virtues) associated with earthen architecture, noting how they are changing and changeable, and linked to political, economic, social and cultural contexts. The impact of these negative and positive values is considerable, affecting practical, archaeological and conservation responses to the material, and involving wider theoretical issues.

Archaeological impact
"Prehistoric architecture did not amount to very much, for if it had, the buildings themselves would have told us such a lot about their builders that the times would be historic." (Williams-Ellis 1946, 37).

The associations and values of earthen architecture have influenced the nature of archaeological interpretation of the past. For example, the linkage between earthen architecture as a cheap, available, and easy to use material, has influenced the type of structures it is associated with. Monumental and high status structures tend to be differentiated not just by design and form, but also by the materials used in construction, with elite or religious authorities having the power and finance to support quarrying, transport and manufacture of building materials. Even from the 4th millennium BC the mudbrick temple structures of Mesopotamia had fired brick, or decorative ceramic cones used for faces and facades, implicit within this is the connection between state-making, kingship (or religion) and notions of permanence, as manifested within buildings (Campbell and Pyrce 2003). This perhaps distinguishes the elite not just through the use of 'prestige' building materials, but also through a disconnection between repair and maintenance (although they can afford, and have the apparatus to carry out maintenance).

Given the physical properties of earthen architecture, these early fired brick and stone structures survived over the longest period in a more complete condition than earth structures. As they are associated with monumental and elite activities it is these structures that attracted the attention of early travellers and archaeologists, as the early evidence of true civilisation:

"The first buildings were very rough affairs, put up from the handiest materials in the easiest way... That kind of almost nest-like building went on for hundreds of thousands of years before, (at last), the highly civilised Egyptians began building to a set pattern in a regular architectural style that could not be possibly be the work of any animal but man." (Williams-Ellis 1946, 37).

As can be read through the above quote, the political context within which these early discoveries took place encouraged the creation of great polarities between the values associated with the surviving different materials, equating stone structures with civilisation, whilst relegating as unimportant the less tangible evidence of our earthen architectural legacy.

Whilst much of the archaeological debate has moved on from the antiquated associations of 'civilisation', it is striking that the poor understanding of building materials, and the negative view of earthen buildings, still creeps into archaeological debate. For example in discussing the development of cities and states in the 4th millennium in Mesopotamia Chris Scarre comments:

"The large agricultural communities that developed in the fertile lowland plains were rich in plant productivity but poor in several other essential materials. It is striking for example, how Mesopotamian cities were built of mud-brick (and occasionally baked brick) but used very little stone...hard stone had to come from the surrounding uplands." (Scarre 2005, 197)

Reading this statement with the sensitivity acquired by my assessment of the positive and negative values of earthen architecture, it is possible to see that in the authors mind 'hard' stone must be somehow better that 'soft' mudbrick. Implicit within this statement is the equation between earthen architecture and desperate action rather than an equation between the use of earthen building materials and environmental suitability. This is a remarkable example, showing considerable misunderstanding of building materials continues even within the most recent archaeological discourse.

Conservation impact
In some instances the negative values associated with earthen architecture influenced the archaeological interpretation and conservation approaches. For example, the excavation at Great Zimbabwe produced a variety of different structures, such as the great stone monuments, but also earthen (*dhaka*) structures. In the atmosphere of

the colonial and racist interpretations of the past at the turn of the twentieth century, many of the *dhaka* structures in the Great Enclosure were destroyed in order to suggest the site was not built by Africans and to link past civilisations with the stone-monuments rather than the earthen structures (Ndoro 1994).

The erosion of earthen architecture is linked to a number of different factors, but generally takes place gradually, punctuated by more dramatic events of collapse. In this respect the erosion and deterioration of earthen architecture is in no way different to other building materials; what makes it different is time. It may take several thousand years more for stone or fired brick materials to erode in comparison with earthen architecture, and when they do erode they may leave more trace, and not be transformed and recycled in the same way as eroded and eroding earthen architecture. The generally negative perception of earthen architecture has also influenced the choice of conservation approaches to the material. Often conservation (and repair in 'living contexts) has been concerned with the utilisation of replacement, notionally harder and more long-lasting materials rather than earthen materials (*Chapters 8-16*).

Living contexts

It is within living contexts that we see perhaps the most dynamic expression of the values and changing values associated with earthen architecture. For example, the revival of earthen architecture in Western Europe and elsewhere is problematic, as the cost of new earth construction in capitalist societies is associated with the costs (and dearth) of skilled labour. This has resulted in the shift in the developed world from the perception of earth as a cheap building material to one associated with elite or specialist activities (*pers. comm.* H. Schroeder; discussions Leipzig 2004). The changing nature, value and perceptions of earthen architecture in Western Europe is of significance, as these changed values, and use of earth in high-profile modern building projects raises the profile of the material throughout the world.

The contextual basis of the associated values of building materials is powerfully illustrated in relation to the Gando Primary School in Burkino Faso (Architecture for Humanity 2006). Here mudbrick was used as a roofing material in order to achieve a passive solar design. Although the mudbrick was used in an innovative way, the use of this local, traditional material was met with initial disappointment:

"According to people in my region, Europeans use more solid materials, like concrete or steel, when they build house for themselves. This is progress"... "But Europeans suggest a different solution for Africans: Africans should keep living in their small, dark clay huts. The villagers found that unacceptable because they equate clay with backwardness." (Architecture for Humanity 2006, 254).

As can be seen what we feel and how we think about different building materials is particularly complex. The Gando example shows how in developing countries the values associated with earthen architecture are particularly complex. Reflecting both 'personal' values affected by the contrast between traditional lifestyles, and lifestyles broadcast and advocated through media images, alongside 'group' values manifested within governmental policies concerned with the expression and assertion of modernity as reflected in capital building projects, and legislation. Commenting more broadly on the state of vernacular architecture and modernisation in the developing world, Oliver (2003) states:

"To a great extent the solutions to immediate housing needs are to be met by national governments and local authorities, engineers and builders, many of whose attitudes and values have been shaped in Western moulds."... "Standardized in plan and structure, multiplied and arranged in geometric settlements for ease of service runs and drawing-board formalism, such bureaucratized, centralized mass-housing solutions are not designed to be responsive either to the cultural patterns of established traditions or to emerging aspirations. They neither utilize local skills nor make intelligent use of received knowledge." (Oliver 2003, 252).

Summary

How and what we feel about a building material is perhaps always shifting and altering, dependent on our concept of 'home' but also societies and cultures conception and expression of who we are. As a result all building materials have fluctuating values and associations. It is possible to understand that all materials embody different values and associations, each creating its own 'aesthetic strand' (De Botton 2006, 195-199).

An example can also be made of concrete, which (like earthen architecture) has an intrinsic value that is forever being re-assessed. Though introduced by the Romans, concrete was only re-invented during the nineteenth century, at this time Victorian society looked on the material with some suspicion, and it was used for bridges but not for 'proper' architecture (Bedell 2005, 171). The enormous use of material in European post-War reconstruction and the campaigning voice of John Betjemen has meant the material has been associated with "ideas of urban, industrial modernity and hardheaded socialist brutalism." (*op cit*, 173). It is only more recently that concrete has enjoyed a growing acceptance, perceived as a quality material in high-end buildings (*op cit*). What this and the proceeding discussion indicate is that the feelings aroused by, and the intrinsic value and associations of building materials are dynamic.

Our associations of and relationship with building materials are complex, affected by both the physical properties of the material (type of construction, ease of use, patterns of retention and deterioration) alongside our own social and cultural perceptions of what a building should look like, and be made of. In this context the perception of earthen architecture is particularly dynamic and complex.

Chapter 5: Problems with earth – difficulties, physical constraints, conservation theory

The understanding of the pathology of earthen architecture provides an awareness of its maintenance needs and conservation requirements, enabling reflection of the physical properties of earthen architecture in relation to current conservation practice and theory. There is a distinction between the active factors which result in gradual loss and erosion and those factors which result in more rapid loss of earthen architecture. These different factors create and pose very different problems for the conservation and management of earthen architecture.

" The houses last as long as they are pleased to repair them, the dry and clean Air contributing to their Preservation; but as I have observ'd elsewhere, the Persians do not like their Parent's Houses, they love to build some fit for themselves, which is very rational; for, as they say, there is the same difference, between building a House fit for one's Family, or taking one ready built, as between making oneself a Suit of Cloathes, or buying one ready made." (Sir John Chardin 1677, 264).

John Chardin suggests the complexity of factors that disrupt maintenance cycles of earthen structures, and lead to the eventual redundancy and abandonment of tructures. Redundancy occurs when a structure no longer retains its use value and is no longer needed. Abandonment is the process by which a structure is given up. Both redundancy and abandonment result in the disruption of the life cycle of a single structure, single compound, entire village or urban zone. Abandonment is the method by which things are left behind, and as a primary form of discard is fundamental to understanding how earthen architecture is transformed from living contexts through to archaeological contexts.

Past work concerned with abandonment has been carried out in rural, semi-rural and village locations: for example, Horne (1993) studied settlement shift in north-eastern Iran; Rothschild *et al* (1993) studied abandonment of settlement in the American Southwest; others (such as McIntosh (1974) and Beazley and Haverson (1982)) examined the general factors that resulted in the abandonment and subsequent deterioration of traditional earthen buildings. All of these studies indicate the complex nature of abandonment; indicating its different nature (both planned and unplanned); causes (related to the life cycle and disruption to the life cycle of a structure); stages (such as the gradual shift from permanently occupied to semi permanently occupied, often as a result of changes in use); and effects on site formation and loss.

Causes of abandonment

The multiple causes of disruption to the life cycle of a single structure, single compound, entire village or urban zone can be seen as those human-related factors, and those associated with natural or environmental factors. In most instances these different factors merge together making it difficult to ascertain a single cause for lack of maintenance and abandonment. Some of the factors may be linked more directly to single structures (death, marriage and laws of inheritance) while others may be associated with entire villages or urban zones (such as ease and access for building).

In summarising these causes of abandonment, the following may be of importance:

- Erosion, decomposition and bio-deterioration of plant materials, for example roofs are recorded as being replaced every 12 years in the village of Abianeh (Iran) associated with the decomposition of the roofing materials (Dekhordi 2004).
- Social and cultural factors, such as inheritance, death and marriage.
- Redundancy of buildings associated with activities no longer carried out.
- The low cost of building anew, alongside the abundance of building materials, and/or availability of land.
- Conflict that results in sudden population movement.
- Economic factors at local, national and international levels that make maintenance costly and/or encourage new construction elsewhere.
- Political factors, ideologies and regimes at local, national and international levels which may envisage and create a disconnection with the past
- Social and cultural change, such as the complex factors that alter the values and meanings associated with a structure or settlement, such as changed perceptions and interpretations of modernity.
- Environmental change resulting in a change in lifestyle, such as desertification, or inundation.
- Environmental catastrophe causing total destruction, and subsequent post-disaster over-work that limits the resources available for repair and maintenance (Beazley and Haverson 1982).
- The relative ease of re-use or adaptation of a structure, in contrast to structures that prove physically difficult to adapt to new uses (such as industrial structures) that may be abandoned.
- The social and cultural perceptions of re-use or adaptation.
- The values and associations of structures and zones, which may make re-use and adaptation inappropriate and unlikely.
- And for earthen architecture in particular:
- The relative ease of recycling and re-use of earthen building materials, for example it may be cheaper and easier to recycle the material and rebuild a structure that has fallen beyond a certain state of repair, such as when roofs fail. The regularity with which structures in the past may have been repaired to a point, and then wholly rebuilt is confirmed by the archaeological

information concerned with tell site formation (see Rosen 1986; Horne 1994).

- The relative ease of building using locally available materials and knowledge.
- The perception of earthen architecture, where negative associations may encourage abandonment and replacement with new construction utilising 'modern' materials such as fired brick or cement (see *Chapter 4*)

A number of studies of the processes of abandonment have observed and discussed different types and causes of planned and unplanned abandonment visible through ethnographic and archaeological research (Cameron and Tomka 1993; Rothschild *et al* 1993; Horne 1993). Planned abandonment is a gradual process that is often linked to the longevity of a structure, such as the decision to rebuild a building that falls beyond the state of reasonable repair. For example, in studying these patterns in Zuni Pueblo (USA), Rothschild *et al* (1993), distinguished patterns of behaviour indicative of planned abandonment, such as the stockpiling of building materials for re-use in new construction. In these scenarios the decision to abandon a structure is influenced by the effort that will be expended on repair when weighed against the availability, ease and costs of rebuilding, alongside other factors such as social and cultural aspects of inheritance and new construction. In addition not all of the structures of a village or settlement may be treated in the same manner, particularly so in traditional societies where family linkage may be evident between many structures and complexes in a village or urban zone (Horne 1983, 1994).

By contrast, unplanned abandonment can be seen as a much more rapid process, caused by the unexpected disruption of the life cycle of a structure as a result of natural, environmental catastrophe or political, social or economic factors. For example, abandonment as a result of earthquake damage or conflict.

The process of abandonment is complex, variable and multi-staged, with one factor affecting another. For example, the abandonment of Zuni Peublos in the Southwest USA was influenced by federal policy that altered land use at the turn of the 20th century, as a result the land failed to support the community, and traditional ways of life and settlement were altered and abandoned (Rothschild *et al* 1993,124). Horne (1993) noted the social-cultural inheritance patterns associated with the traditional *qanat* system in north-eastern Iran, identifying how through time, lines of inheritance multiply and become too numerous for sufficient profits to be generated to allow for *qanat* maintenance. As a result the man-made environment and its capacity for supporting agriculture fluctuates. In response settlement patterns would alter, sometimes resulting in redundancy and abandonment (Holmes 1975 *cited* in Horne 1993). Factors such as these are associated with the abandonment of Iranian settlements such as Shahdad (Iran) (*Appendix 6*). Here the fortified *qala* structures associated with those local rulers, who previously retained control over *qanat* systems, have been abandoned as a result of changing cultural, social and economic patterns. Within Shahdad (Iran) the effects of this abandonment have been amplified by the relative difficulties associated with the re-use and adaptation of such large *qala* structures for current uses, alongside more rapid disruption to the rural population associated with earthquake damage.

In other instances buildings are abandoned due to redundancy; they no longer serve the original function, and remain unused for other purposes. For example, a levelling of social stratification and change in social and economic patterns may result in high-status buildings becoming obsolete, such a pattern is seen with the *qalas* in Iran, Turkmenistan and Uzbekistan (Horne 1993; Hallet and Samizay 1980; *pers comm.* Gaigysyz Juriev). Other monuments associated with religious, elite or commercial activities were abandoned within the former Soviet controlled regions of Central Asia as a result of changing political ideologies that forced redundancy. For example, the Maddrassah at Taklahtan Baba (Turkmenistan) was abandoned due to changing religious uses, and latterly the suppression of Islam in Soviet Central Asia (*Appendix 6*). In other instances the function of a structure may become obsolete such as icehouses (made obsolete by electrical refrigeration) and pigeon towers (made obsolete by changing agricultural practice) (Beazley and Haverson 1982). Because of the original form, the adaptation of these sorts of structures may be particularly complicated, and often obsolescence coincides with other changes such as rural depopulation. In these cases the disruption and abandonment results in the transformation of these structures from living contexts through to abandoned and archaeological contexts.

Worldwide, patterns of redundancy of historic towns, transforming structures from contexts of maintenance through to contexts of abandonment and deformation can be seen. In contexts where these processes of transformation are currently active, the abandonment of the historic towns and structures has been encouraged by the perception by both governments and individuals of the un-adaptability of historic towns to a modern infrastructure incorporating wide roads and drains, alongside difficulties of adapting traditional structures to modern needs, such as the incorporation of electricity and sanitation (working group discussions Terra 2003). The result of these changes is the depopulation of traditional housing zones, and the abandonment of maintenance regimes, resulting in increased rates of deterioration. Studying the historic town of Shibam (Yemen), Damluji (1992) classified the reasons why populations have shifted from the inside to the outside of the historic zone, as a result of the requirements for space to construct, rebuild and repair structures; the cost of bringing in repair materials to the crowded historic city; the need for deep foundations for safe construction in the historic city; and construction laws regarding the retention of the character of the historic town, these factors combine to make construction within the historic town more expensive and perceived to be less well suited to modern needs (Damluji 1992, 191). Similar examples have been seen worldwide, and within the study area. For example, the old town of Yazd (Iran) has been gradually abandoned since the 1970s, as a result of complex and changing economic and social perceptions of modernity and habitation (Khademzadeh 2003, 2004) (*Appendix 6*). Whilst the old town of Bukhara (Uzbekistan) was partially, and forcefully abandoned in the 20th century, as a result of disruption caused by the political, social and economic idealism of Sovietism in Central Asia (Gangler, Gaube and Petruccioli 2004) (*Appendix 6*). It is estimated that in Bukhara the disruption of traditional interactions in the historic town resulted in the movement of half of the population of the old town (*c.* 40,000 people), and its re-housing in prefabricated Microrayin blocks located outside the historic town (*op cit*). In contrast policies for the retention of the historic town of Khiva (Uzbekistan) resulted in its sanitisation through the forceful removal of the local population in order to create an idealised museum city (*Appendix 6*). In Khiva (Uzbekistan) this complex process of abandonment resulted in the retention of the high status monuments associated with religious or elite activities, and with the loss of the traditional structures that comprised the urban infrastructure and texture of the old town.

The processes of abandonment are also affected by the changing perception of building materials, and associated skills and craftsmanship (see *Chapter 4*). The importing of building materials in areas formally utilising earth building materials has altered labour markets, and facilitated low-cost, relatively unskilled labour in construction. As a result the skills associated with earthen architecture are in less demand, and require few craftsmen and apprentices. This means maintenance, repair and construction activities decline and/or the costs of employing those specialised craftsmen becomes too great for all but the wealthiest in society, meaning it becomes much more complex and complicated to repair traditional earthen buildings and plan for the adaptation and re-use of earthen buildings. This results in an increase in inappropriate repairs, abandonment and replacement with new construction utilising non-earthen elements.

Manifestations of abandonment

The different reasons why buildings fall out of use result in different responses to the remaining building fabric, and just because a structure is no longer lived in, it does not mean that it is abandoned (Rothschild *et al* 1993). Redundancy of a structure may result in a transition from permanent to semi-permanent occupation, for example living spaces may in time become *re-used* semi-permanently for seasonal occupation (*op cit*). Similarly, Horne (1993, 1994) notes the complex and variable stages in which structures change use, noting how over a period of time human spaces may be transformed into animal or storage spaces (but rarely *vice versa*). These types of responses to the changing values associated with structures, and the transition from permanent to semipermanent occupation are important in understanding the process by which earthen architecture is formed into archaeological contexts. Associated with many of the examples of changing use, and/or semi-permanent occupation is less emphasis on maintenance. In these instances there may then be a transition from permanent to semipermanent to complete abandonment. For example, in Konya (Turkey) fragments of the historic urban fabric remain but within the modern city the values and associations are transformed, as a result the historic structures are marginalized against the modern urban backdrop, where they are retained and used as ancillary structures, such as garages and boundary walls for car parks (see site dossier *Appendix 6*).

Important in understanding the lifecycle of structures is the fact that these transitions may be reversed at any point, for example reverting back from semi-permanent to permanent occupation, or from unoccupied to occupied contexts. These processes change the values associated with a structure and practical approaches to re-use and maintenance. For example within the old city of Yazd (Iran), many of the abandoned structures are re-used by squatter refugee communities. These communities have less economic capacity for maintenance, and are more familiar with different forms and uses of earthen architecture (such as flat timber and earth roofs, rather then domed mudbrick roofs) (*Appendix 6*). Similarly, a family has re-occupied the abandoned complex at Taklahtan Baba (Turkmenistan), re-using a range of the courtyard structures, and adapting them to fulfil their present day requirements, such as with the repair and rebuilding of the domed mudbrick roofs with cement and fired brick/tile (*Appendix 6*). These instances of re-use result in a transformation of the approaches to maintenance, alongside changes to the values associated with the structures.

The eventual abandonment of a structure may also result in the re-establishment and replacement of a structure on the same piece of ground. A structure may be replaced following on from planned abandonment, when a building has fallen beyond the point of reasonable repair, in these cases the building may have the same use and values associated with it as the building that it has replaced. This pattern of replacement has been used to interpret the formation processes associated with earthen architecture retained on archaeological tell sites (Horne 1994; Rosen 1986). However, this simple interpretation underinterprets the complexities associated with abandonment and subsequent site formation, for example the pattern of replacement at Çatalhöyük (Turkey) is much more complex with use, abandonment and replacement varying spatially and temporally across the site (*pers comm.* Shahina Farhid).

The eventual abandonment of a structure may also result in the re-establishment and replacement of a structure within a new space. This type of removal may be planned with the stockpiling or re-use of robbed building materials from the abandoned location. In understanding the processes of abandonment associated with earthen architecture, it is important to consider the manner in which abandoned components of earth buildings are transformed by the ability of the material to be fully recycled. As a result structures which have suffered extensive erosion may be reworked, with the materials of the earth walls broken down and re-used either in the same form of the original construction - mudbrick as mudbricks (Damluji 1992, 130) - or in a different forms of earthen construction - mudbricks recycled as placed earth walls. For example Horne (1994) noted that earth building materials came from locations in and around villages where earth from the abandoned and eroding older structures was reworked into *chineh* walls (Horne 1994 130). Similarly at Merv patterns of recycling are interpreted within the archaeological and contemporary evidence for earth building (Puschnigg 2001; Cooke 2004). Here material from the older eroding buildings is sought out and recycled for new mudbrick manufacture, as the already worked material is easier to use (*op cit; Chapter 6*). The re-use of building materials may be indicative of both practical and pragmatic responses to relocation and new construction, such as the scarcity of available building materials, alongside more symbolic acts of re-incorporation of the old and with the new, through the re-use of earthen building materials.

In this context it is also of use to consider the social, economic and political factors that result in the survival of structures in contexts of abandonment. A final result of abandonment is the transformation of the abandoned place and site, which may still be visited, seen and remembered, acquiring new values and associations through time. At Merv some of the structures, such as the larger *khöhsks* survive exactly because they were abandoned, and were not reused or rebuilt after the Mongol invasion (Herrmann 1999). At Merv the fact that the settlement has shifted across the landscape through time has resulted in the retention of the archaeological evidence of the historic cities. Similarly perhaps the final stage to consider is the process by which abandoned structures are singled out within contexts of conservation for adaptive re-use and restoration, or for retention as monuments.

As will be seen in the following section, once structures have been abandoned and transformed into contexts of deterioration a number of taphonomic processes will transform earthen architecture from abandoned but upstanding structures, to the formation and deformation of earthen archaeological deposits.

Erosion and deterioration

The following section discusses the erosion and deterioration factors that effect earthen buildings once they have fallen out of use. The effect of erosion and deterioration is the transition from used, to abandoned but upstanding structures, to the formation and deformation of earthen archaeological deposits. The type of erosion and factors that cause deterioration occur to earthen architecture in all contexts, however these damaging effects are checked and managed in contexts of use through patterns of regular maintenance and corrective repair (see *Chapter 6*). It is when maintenance ceases and these structures are transformed into contexts of abandonment that the effects of erosion and deterioration accumulate and multiply.

The types of erosion and deterioration factors include decay factors inherent in the properties of earthen architecture, environmental damage and human agency. These factors rarely occur in isolation and rather work in combination, with one factor leading to the next, magnifying the effect, and so on gathering momentum (Hughes 1988). Earthen architecture exhibits a non-linear pattern of deterioration, and the types, degree and extent of erosion to earthen architecture are affected by the local environment, in particular annual rainfall, humidity and moisture, and annual and diurnal temperature variation. As a result whilst a broad class of factors that cause deterioration can be assessed, the impact of each of these factors will be context and site specific (Sketch 1-4).

Gradual loss and erosion	**Catastrophic or rapid loss**
Wind erosion	Earthquakes
Falling water	Fire
Rising water	Floods
Visitor damage	War
Bird nesting	Development threats
Re-use of materials	Re-use of materials
Vegetation	Saturation
Burrowing	Unusual weather patterns (such as El Nino)

Table 2. The types of gradual and rapid loss effecting earthen architecture.

Some the types of erosion and deterioration factors result in catastrophic and rapid loss, and some result in more constant and gradual losses (Table 3). The length of time in which a structure has been eroded and eroding or has been an archaeological site is important in understanding the nature and type of damage to earthen architecture. These time-based factors are illustrated through the comparison of the condition of a selection of the monuments at Merv recorded in the historic photographs (*Appendix 5*). The longterm effect of the continual loss and erosion will be the gradual formation and deformation of earthen archaeological deposits. This gradual process is punctuated by more dramatic episodes of loss and collapse. For example, the continued attrition and erosion of the Great Kyz Kala at Merv eventually led to the more dramatic loss and collapse of the central section of the eastern wall, reducing the number of corrugations on this side of the monument from 22 in 1890, to 16 in 2003 (Fig 59-60; *Appendix 5*).

Effects caused by materials

As identified in *chapter 2*, soils comprise mineral and organic components, which derive from the decomposition and weathering of parent materials (Limbrey 1975). The key components of a soil that influence its suitability for use in earth construction are the mineral components and in particular the type and characteristics of the clays present. Clays are natural aluminosilicates and can both absorb and adsorb water, whilst variations in temperature result in expansion and contraction. The susceptibility of clay (and the different clay types) to swelling and shrinkage results in the destruction of the basic mechanical bonds in the material, as clay acts as the binding material in earthen architecture this gives rise to shrinkage, cracking and shearing (Balderamma and Chiari 1995; Brown and Clifton 1978; Brown *et al* 1990; Torraca 1981). As a result the qualities of the original materials used in construction have an enormous effect on the longevity or otherwise of earthen architecture. For example in desert environments the calcite rich materials that may be within the soil used in construction can contain a high quantity of calcium carbonate. This natural consolidant imparts strength to the substructure as through its weathering and chemical breakdown, it fills up the pore spaces in the matrix (see Limbrey 1975; Kemp 2000). The mechanical weathering of the original materials also accounts for the great strength and resistance imparted through time to the chalk structures in the United Kingdom (Pearson 1992).

Given the great variety of different earths, and earth mixes used for earth construction, the different earth construction elements of a structure (mudbricks, mortars and renders) will have different properties, and differing resistance to erosion. As a result a structure may suffer from differential erosion (Hughes 1983). 'Honeycombing' may occur in a mudbrick building where the masonry erodes leaving behind the earthen mortar, or 'reverse honeycombing' where the earthen mortar erodes leaving behind the masonry (Fig. 65-66). Similarly in rammed or placed earth construction the different properties of the earthen building materials can result in differential erosion of the separate lifts (*pers comm.* Archie Walls; Fig. 40, 82). Whilst the different properties of materials used in repair may be less durable and erode more quickly, conversely they may be more durable and induce erosion of the surviving fabric (Hughes 1983).

The inclusions within the earth mix used for construction will also affect the longevity and survival of the structure gradually through time. The decomposition of plant material within the earthen matrix is a factor that determines survival or deterioration, with the resistance to cracking of an earthen structure decreasing through time, associated with the decomposition of the straw bonding material (Oates 1990). The decomposition of plant material and the loss of its benefits, alongside the creation of voids within the earthen matrix can place the material under much greater stress (Hughes 1983). For example, at Çatalhöyük (Turkey) one of the problems apparent with the conservation of the wall paintings exposed through the course of excavation is rapid desiccation and drying out, resulting in shearing of a zone that is equivalent to the mud plaster render coat. This shearing may be associated with the decomposition of the plant material (*pers comm.* Wendy Matthews). Problems associated with the decay and decomposition of organic materials has also been seen to be one of the factors linked to the loss of the historic fabric in the Bam (Iran) Earthquake (Vatandoust and Mohktari 2004). Similarly as mudbricks with a high straw content may be preferred for use in arches and barrel vaults (Fathy 1973), the decomposition of straw in these bricks may prove particularly problematic, and may account for some of the more characteristic patterns of erosion associated with earth roofs (see below).

In other instances the erosion, weathering and deterioration of earthen architecture may lead to the leaching out of the clay content through time (for example with the continued washing away of fines). This is indicated as the materials analysis of archaeological and historical earthen architecture often indicates a reduced clay content in historic and archaeological materials, than that recommended or evident in new earth structures. The loss of the clay-binding agent may account for the increased propensity to shearing, cracking and crumbling associated with earthen architecture that has been exposed over a long period of time (discussion Merv).

Environmental damage

Environmental factors, such as falling water, wind erosion and freeze-thaw damage, alongside animal, bird and insect activity, and vegetation growth result in damage to earthen architecture (Hughes 1983). The surface damage and alteration caused by these environmental factors is moderated in contexts of use through human action and the processes of maintenance and repair. However for earthen architecture in contexts of abandonment, this damage is not moderated, and can exponentially grow, with the overall loss of definition and surface damage leading to much more extensive deterioration and loss. Most environmental damage is affected and moderated by the orientation and exposure of the structure, giving rise to differential patterns of erosion(as at Merv, *Appendix 5*). These differential erosion patterns can in turn stress and alter the load bearing characteristics of the structures, again exponentially increasing the deterioration and loss of structures.

Water

The various different effects of water cause most substantial damage to earthen architecture. In the same way that water must be used to convert the dry earth into a soft, malleable material to enable earthen construction, water can be thought of as the 'activator' for erosion. The clay component of any earthen building material can be converted to a liquid form in the presence of water, which is then liable to being washed out, or to shrink and expand in relation to thermal variation (Hughes 1983). Water related damage also occurs due to rising water, precipitation, and freezing water; and in some cases there may be more rapid losses associated with the saturation of earthen architecture. In addition to encouraging mechanical loss and capillary rise, increased moisture content and absorption will generally decrease both tensile and compressive strengths of earthen architecture (Brown and Clifton 1978; Balderamma and Chiari 1995; Hughes 1988).

Rising water
Rising water damages the base of earth buildings. Capillary action occurs in any porous material, and causes moisture to rise and spread (Torraca 1981). When the moisture reaches the surface it evaporates, and the clay component shrinks. The absorption of groundwater with an abundance of soluble salts (and agrochemical residues) exaggerates the effects of capillary damage. If the rising water contains soluble salts and these are transported in the moisture by capillary action, these salts can crystallise out of the moisture either at the surface (efflorescence) (Hughes 1983, 183) or sub-surface (subflourescence) (Balderrama and Chiari 1995, 102). The effect is to break down the bonds between the materials and cause the exfoliation and the detachment of the surface crust. The result of this sort of damage results in characteristic erosion of the wall base and earthen structures - basal erosion and/or undercutting (Warren 1993).

Different types and degrees of undercutting indicate different rates of loss (or causes of loss). From assessing the damage that has occurred to the monuments at Merv, it can be seen that this type of damage is accelerated when undercutting occurs at corners, where two already undercut horizontal profiles join, and this extends the zone of damage vertically up the wall (*Appendix 5*).

Damage to the wall base is exponential, with the resulting deposits of loose, spalled material drawing the evaporation zone up through the surviving fabric, and raising even further the area susceptible to damage. The erosion mounds that form from the eroded and eroding earthen architecture will also raise the zone of evaporation, making erosion of the wall base migrate up the wall with the height of deposition of the erosion material and mound. This phenomenon is recorded on the southern wall of the Little Kyz Kala at Merv where, as a result of mound clearance, a zone of undercut damage is now visible above an older zone of undercut damage (*Appendix 5*).

As buildings become progressively more damaged from undercutting at the base they may lose stability, and may collapse as a result of changing loads and stresses (Hughes 1983). Since earthen architecture is typically strong in compression but weak in tension (Walker *et al* 2005), if the base of an upstanding earthen building becomes undercut, this means its compressive strength is reduced and load-bearing characteristics altered. Such a phenomenon can be identified associated with the collapse of the extant walls in the Palace in Shahriyar Ark at Merv. Here construction employed much thinner mudbrick walls in comparison with the great mass employed for the corrugated structures at Merv, and as a result undercutting reduced the capacity of the walls to withstand alternating dynamic wind loads resulting in eventual collapse (Fig 67 & 68; *Appendix 5*).

Falling water
Falling water in the form of rain or snow damages earth buildings, initially eroding the exposed surfaces, through the expansion and mechanical weathering of the clay component of the earthen materials (Hughes 1983). Surface run-off results in damage through the reworking and washing out of the fines from the horizontally and vertically exposed surfaces of upstanding earthen architecture and earthen archaeological sites (*op cit*; Walker *et al* 2005). Surface run-off causes the continual washing and movement of the fines from the surface with a subsequent loss of definition. In contrast surface runoff can also give rise to surface capping, and crack infilling, as the washed out fines accumulate on the surface or in voids and cracks. The continued erosion, deposition and redeposition of earthen material alters the appearance of the exposed surfaces, and erosion mounds, but rarely is the single contributing factor to more dramatic, structural loss.

The long-term effect of the washing and movement of earthen material is the obscuring and burial of the surface. Surface run-off also results in the thinning and shortening of upstanding earthen structures (the decrease in height and width at the wall tops), and shortening and widening of the earthen archaeological sites. In some instances this sort of washing and redeposition by surface run-off has buried the lower deposits of structures and buildings, assisting survival as archaeological tell sites (McIntosh 1974; Rosen 1986). These continual processes of deposition and redeposition caused by water run-off alter the shape and form of earthen structures and archaeological sites, leading over a very long period to eventual loss.

More concentrated water flow over the exposed surface creates runnels, and larger gullies (channels worn by running water). The formation of gullies is particularly apparent in upstanding earthen structures in areas of already imposed weakness, such as putlogs, windows, or doorways; on archaeological sites in those areas where the ground surface crust is broken; and in areas damaged by excavated nests (Hughes 1983, 1988). These zones have a tendency to initiate the formation of gullies through the channelling and redirection of water flow over the face of the structure (Peek 2004). For example, the study of the patterns of erosion visible on the structures at Merv indicates the propensity for run-off gullies to form in the location of voids (windows, entranceways and stairwells), on the eroded and eroding earth structures. These gullies channel water down the face of the structures, whilst the zones immediately surrounding the void will become thinner and the voids become wider and higher through abrasion, erosion and collapse of fallen or loose material. Similar effects of localised damage and erosion are also seen on the corrugated buildings at Merv, where extensive gullies have formed at the base of the corrugations, in the zone of transition with the plinth on the Kepter Khana at Merv (Fig. 61 & 62; *Appendix 5*).

Falling water at wall bases also contributes to a splash zone, where if the ground does not absorb excess moisture it is splashed back onto the vertical surface (Hughes 1988). This excess moisture at the wall base can accelerate erosion in a zone that is already susceptible to increased erosion from undercutting action.

Freezing water

Damage caused by excess moisture is exaggerated in freezing conditions. This is because frost damage can open up the microscopic air spaces in clays, and moisture absorbed by the clay component of the structure will expand when it is frozen, and damage even further the bonds between the materials (Hughes 1983). Damage is particularly problematic when snow sits on top of the surface for a long period. The weight of snow on the roof of an earthen structure is a static load that may take time to slowly accumulate and alter the loading characteristics of a structure. The rapid thawing out of this frozen water will further saturate the earthen materials, and rapid surface runoff will encourage the formation of drainage gullies.

Wind

Wind causes deterioration of earthen architecture through the accumulated effects of abrasion and through the alteration of dynamic loads (Hughes 1983). If wind carries sand particles it abrades exposed surfaces and results in a substantial loss of material (Balderamma and Chiari 1995). Wind can also increase the speed of, and encourage evaporation to take place immediately below the surface of the exposed deposits (*op cit*; Brown *et al* 1990). The drying out of the surface caused by wind can result in excessive exfoliation. The extent of wind erosion is determined by local and micro- climates, and by the setting and orientation of the structure. For example the corrugated earthen buildings at Merv are most eroded on the northern faces, whilst the southern faces are best preserved. This has resulted in differential preservation on the interior and exterior faces of monument, with the interior of the north walls of monuments remaining well preserved, in contrast to the poorly preserved exterior of the north facing walls. Such a phenomenon is recorded on the Kepter Khana at Merv, where the northern and western sides of the monument have suffered greater loss through abrasion resulting in the thinning, shortening and gradual loss of definition and distinction between the plinth and the corrugations (Fig 63 & 64; *Appendix 5*).

Wind also causes deterioration to earthen architecture through alteration to dynamic loads. Consequently loose parts of upstanding earthen structures can become detached and, in extreme cases wall loads redirected (Hughes 1983, 183). This is because wind acts dynamically on a structure, altering and redirecting the aerodynamic characteristics of walls. This can give rise to failure and/or collapse of elements of the upstanding earthen structure. The long-term effect of wind on earthen archaeological sites includes 'tell-creep', whereby wind erosion (with surface run-off) manifest in the deposition and redeposition of earthen materials, and the movement of tells in line with the direction of the prevailing wind (Rosen 1986).

Vegetation

Vegetation is problematic for upstanding earthen architecture and earthen archaeological sites through invasive roots (Warren 1993, 1999), and through the creation of microclimates. Plant roots can cause damage in a radius up to 2.5 times the height of vegetation (Hughes 1983). Vegetation can grow through upstanding earth walls and earthen archaeological sites, resulting in physical loss and alterations in structural stability. Vegetation can also trap moisture, raising the relative humidity and lowering ambient temperatures (*op cit*).

Different types of vegetation effect earthen architecture in different ways, desert plants are particularly problematic as they have extensive root networks. For example the most common vegetation at Merv is an invasive self-regenerating perennial camel thorn (*Alhagi. sp*), this has an extensive root network that can cover an area of up to 15m in area, and at a depth of 2m (Merv Backfilling Report). Other plants may have particular adaptation features designed to retain moisture during the day, such as desert tamarisks (*Tamarix*), these have a large surface area in order to retain a high quantity of moisture, which in response to diurnal temperature fluctuation will condense at night (Fig 76 & 80; *pers comm.* Archie Walls). Through the creation of a microclimate, these plants have the capacity to cause more rapid deterioration of upstanding earthen architecture and impact and damage earthen archaeological sites.

In contrast vegetation growth may also be a factor that contributes to the stability, compaction and survival of earthen architecture. The presence of vegetation at wall bases may be beneficial as by absorbing excess moisture, vegetation can reduce instances of splash back of falling water onto walls, whilst vegetation cover can limit wind abrasion and erosion on archaeological tell sites (Horne 1994). In other instances research has been carried out on tell sites to investigate the best plants to stabilise erosion gullies (Miller and Bluemel 1999). As a result the removal of some plants may actually contribute to erosion and deterioration of the surroundings of upstanding earthen architecture, and the erosion of the surface on earthen archaeological sites.

Animals

Animals, birds, insects, and reptiles cause extensive damage by burrowing through and excavating nests in earthen walls and earthen archaeological sites. Burrowing removes soil, and may eventually result in collapse (Hughes 1983, 183), whilst birds such as swallows recycle earthen materials in nests, and insects excavate and redeposit earthen materials in glassy cellular nests. Animals are attracted to abandoned earthen architecture on account of trapped moisture and lower ambient temperatures (*op cit*). For ease of burrowing and nesting, animals exploit weaknesses that may have already formed in the structure, such as putlogs, and zones of weakness caused by the different and variable nature of materials used. For example, at Merv birds have a tendency to exploit zones where there has been differential erosion of either mud mortars or mud plasters, and voids in which archaeological excavation techniques have been to remove mud mortars in order to articulate and make visible mudbrick courses. These excavated nests then have a tendency to initiate the formation of gullies (see above, Peek 2004).

Birds, animals and reptiles can also alter the chemical composition of the exposed structure by creating an accumulation of loose phosphate-rich soil through food waste, guano and faeces (Hughes 1983). Damage is also caused through abrasion, for example camels have been observed at Merv, rubbing up against the still upstanding, but eroded earthen elements, eroding the walls (particularly at the corners) in a location that corresponds to the height of the animals.

Construction detailing and erosion

Construction detailing, design and form of structures are factors of significance in assessing the longevity or otherwise of earthen architecture and the nature, type and rates of deterioration and erosion (Hughes 1983).

The materials and techniques used for the construction of earthen architecture accounts for some of the erosion patterns seen. For example, the *paksha* construction techniques recorded in Uzbekistan and Turkmenistan account for some of characteristic patterns of erosion and deformation of *paksha* walls, such as cracks forming at the construction seam between the horizontal layers, and diagonally through the construction seem boundaries. In other instances multi-period structures that have been successively rebuilt or repaired in different styles, materials and techniques may suffer differentially at the interfaces between the different phases of construction, such as differential damage from water infiltration as a result of the different properties of the earthen building materials used in different phases of construction.

There is an association between the types of erosion that may occur within structures constructed with different roof types. Such are the benefits of mudbrick vaults, arches and domes in resisting termite attack, in contrast to flat roof structure, these techniques were re-introduced in the course of the 20th century in Yemen as a means to avoid problems associated with deterioration of roof timbers (Damluji 1992). In addition when flat roofed buildings fall out of use the timbers used in construction may be reused and recycled, as a result the superstructure and load-bearing elements become exposed to weather, and accelerated processes of deterioration. In contrast domed or vaulted roof structures that fall out of use may not have roofing materials robbed, and as a result have a longer period of deterioration prior to the collapse of the roof and the exposure of the superstructure and load-bearing elements to weathering. These distinctions may also create differences in erosion mounds and tell formation associated with the two different forms of roof constructions, with flat roof structures (without robbed timbers) forming flatter erosion mounds than domed roof buildings with larger quantities of earthen material creating higher and more extensive erosion mounds (Horne 1993, 169).

The wall thickness and mass of the original structure are of importance in determining the rates, nature and extent of loss associated with earthen architecture, with the thickness of an earthen wall influencing its longevity (discussion Merv). Therefore those structures that are of a greater mass at the time of construction are more likely to both survive upstanding for longer, and form more substantial erosion mounds comprising eroded and buried earthen materials. These more substantial mounds will resist erosion for a longer period of time than less substantial mounds. As mass is such an important factor in the survival or otherwise of earthen structures this accounts for the differential survival of earthen structures on archaeological sites. This is well illustrated by the differential survival recorded at Merv, where the thinner mudbrick walls of the Palace in Shahryar Ark, erode more quickly, and leave less trace than the massive corrugated structures, such as the Great Kyz Kala (see above Fig. 48, 49, 59-68; *Appendix 5*).

One of the characteristic erosion patterns indicative of the type of damage caused by falling water is the gradual thinning and shortening of earth walls and structures. As a result the make-up of the wall will determine its resistance to thinning and shortening.

Damluji (1992) records the manner in which the size and shape of mudbricks varies with height in the traditional tower block buildings in Yemen's Wadi Hadramawt, and *paksha* walls constructed in Central Asia taper in as they increase in height. Therefore, in contexts of erosion and deterioration, the upper storey and/or lifts of these walls will be more likely to erode quicker, and exaggerate the properties of earth walls to become thinner and shorter through time. It is also the case that boundary and non-load bearing walls may be of thinner construction in comparison with load bearing walls, given the relationship between wall thickness and longevity it is likely that the thinner, nonloading bearing walls will erode quicker. The characteristic inverted v-shape of eroded earthen walls, is assumed to be better at throwing water off and away from the main body of the structure, perhaps therefore the shape of the wall tops adds a degree of longevity to earth walls in contexts of abandonment and deterioration (Caperton 1990).

The nature of foundations and wall bases influences the potential and nature of damage to the base of structures through capillary rise (Hughes 1983). Foundations and wall bases utilised for earthen architecture tend to have great variation depending on the local environment and characteristics, but wall bases generally tend to be constructed from stone or fired brick (and today concrete and breezeblock) to a height c.30 – 100 cm. The type and longevity of materials used for the construction of the capillary break influence the long-term survival or otherwise of earthen architecture.

The nature of the wall configuration and supports can also affect the deterioration of earthen architecture. The study of the historic and new photographs from Merv indicates that areas with particular construction details can erode deferentially, for example the chequerboard pattern utilised for the wall construction in the palace in Shahryar Ark (Fig. 48 & 49; *Appendix 5*), and the decorative mudbrick coursing within the Icehouses at Merv, show patterns of differential erosion. These may be associated with differences in the materials used in construction, the resistance of these different materials to factors such as water infiltration and surface run-off, and the pattern and wall configuration that can actually encourage the gullying and collection of water.

Design features such as the corrugations and crenellations can also result in distinctive patterns of erosion and loss by localising run-off patterns. For example, on the reconstructed walls in Khiva (Uzbekistan) and Bam (Iran), water run-off has gathered at the base of the reconstructed crenellations, resulting in the localised washing out of fines from the surface. These patterns can be predicted as leading to the formation of runnels, and more extensive gullies, that may channel and redirect water run-off down the surface of the reconstructed walls, whilst the loss and blurring of surface detail will result in the formation of distinctive waved patterns in place of the crenellations (Fig 75 & 83).

People

Damage by people to earthen buildings may be deliberate, such as the systematic removal and relocation of building materials for re-use and recycling (see above). These types of re-use are particularly associated with elements that are easily recycled, or particularly scarce, such as timbers in a desert environment, or expensive, such as fired bricks. Rates of erosion and deterioration are accelerated when this robbing is of protective elements, such as timbers used in roofs, or fired bricks used for damp-proof courses. In other instances earthen materials from eroded and eroding structures may be recycled as manure for agricultural fields (Damluji 1992). Visitors to earthen buildings or archaeological sites contribute to gradual attrition. This is the result of foot and hand erosion and the effects of altering environmental factors (such as salts and humidity) during visits (*pers comm.* Tim Williams). For example movements by visitors, and vehicle access can break the hardened surface crust on an earthen archaeological site, this layer of loose, and more porous material accumulates on the surface forming an erosive matrix, which during rainfall is liable to wash over the surface, creating erosion gullies and through eventual deposition and accumulation obscuring the setting and context of the archaeological site (Cooke 2002).

Damage also occurs to eroded and eroding earthen architecture as a result of inappropriate repair. For example, measures undertaken to repair (and conserve) buildings that have utilised harder, cement based materials have contributed significant damage to earthen architecture, as these sort of materials create a hard, impermeable barrier below which there is an increased rate of erosion and deterioration.

Damage also occurs to eroded and eroding earthen architecture as a result of archaeological excavation. In the past, excavation strategies were such that they rarely considered the importance of planning well for the trench location and spoil heap location. If excavation trenches are left open, damage occurs in a zone that exceeds the area of the original excavation (Cooke 2002). These open trenches may then be re-used or reoccupied by burrowing animals and/or retain moisture and so increase the zone within which erosion and damage is occurring.

Many of these human related factors associated with rapid deterioration and loss are interconnected. For example the dramatic loss of the western gateway of Abdullah Khan Kala at Merv, recorded between 1903 and 2003, has been a result of human related actions, such as robbing and removal as this once provided vehicle access through to the interior of the monument for rubbish dumping (Fig. 69 & 70; *Appendix 5*).

Local and temporal context of erosion and deterioration

The factors that result in abandonment and all of the erosion and deterioration factors are contextually dependent, and this accounts for the very variable survival of eroded and eroding earthen architecture seen at a global scale. For eroded and eroding earthen architecture the extent of surface run-off, moisture rise and wind erosion are contextually dependent and very local, with variation seen along single stretches of wall, whole structures, and abandoned complexes. On archaeological sites factors such as the depths at which moisture percolates down or capillary action moves up, and the depth of burial at which diurnal and annual temperature variations are limited, are contextually dependent and locally variable. This means that deterioration will vary dramatically across a site, monument and even single wall. As a result whilst general patterns can be postulated the extent and effect of erosion and deterioration is locally dependent. These effects are well illustrated at Merv, where wind erosion is much more likely to flatten out and abrade north facing walls, whilst those structures located adjacent to ancient or modern canals are more likely to suffer damage as a result of excess or trapped moisture at the wall base (*Appendix 5*).

The variable nature of erosion and deterioration also accounts for the importance of approaching earthen architecture holistically, and assessing the importance of a building, monument or site setting and context. Due to the variable, and local effects of deterioration it is important to assess the impacts of erosion on a whole site, and in management contexts, to undertake activities to manage and remedy these factors holistically, understanding how one factor may influence others, and understanding the knock-on effects of conservation and management activities in worsening and acerbating erosion and deterioration factors.

The phenomena that cause erosion of abandoned earthen architecture can be seen as exponential, with one factor exacerbating another. If the abandoned structures are not re-used, quarried for building materials, and not built upon, but rather left as isolated and untouched monuments there are a number of different stages of erosion and deterioration that result in the progression from newly abandoned earthen achitecture, through to eroded ruins, through to earthen archaeological sites.

Sites visited within the study area, such as Yazd (Iran), Taklahtan Baba (Turkmenistan) and Khiva (Uzbekistan), all had examples of recently (*c.* 15 years to present) abandoned structures (Fig. 71-74, 77, 78, 79, 81, *Appendix 6*). These abandoned structures all showed patterns indicative of the rapid loss and erosion of earthen architecture within the first few years of abandonment, as a result of a lack of maintenance, and corrective repair. This included substantial erosion of the wall surface, as a result of the washing out of fines; the creation of erosion runnels, leading to more substantial erosion gullies; and erosion to a zone at the wall bases as a result of capillary action and splash back. These factors lead to roof collapse, which is perhaps the most dramatic stage in the deterioration of earthen architecture. In Yazd (Iran) and Taklahtan Baba (Turkmenistan) the mudbrick vaulted and domed roof structures showed patterns in which they would characteristically crack at the top of the dome or vault, where surface erosion has made this zone thinner, and/or collapse where deep erosion gullies had formed at the transition between the load bearing wall and the dome. In contrast, in Khiva (Uzbekistan) the bio-deterioration of the timbers and reeds used for roofing resulted in collapsed zones.

Following on from the loss and erosion of the roof structures the wall tops become exposed to falling water, which can enter the structures through voids, and become trapped within the loose collapsed materials. At this stage the erosion of the earth structures, again takes on more gradual characteristics, with erosion of the wall surface; the continued erosion of and creation of new gullies; erosion to the wall bases resulting in more substantial undercutting, and erosion to the exposed wall tops.

These gradual processes of erosion may be punctuated, by more dramatic events such as when the remaining extant wall sections experience collapse as a result of the failure of heavily undercut zones, and/or buffeting by heavy winds. Through time these continued processes affecting the eroded and eroding earthen architecture results in the formation and deformation of archaeological deposits.

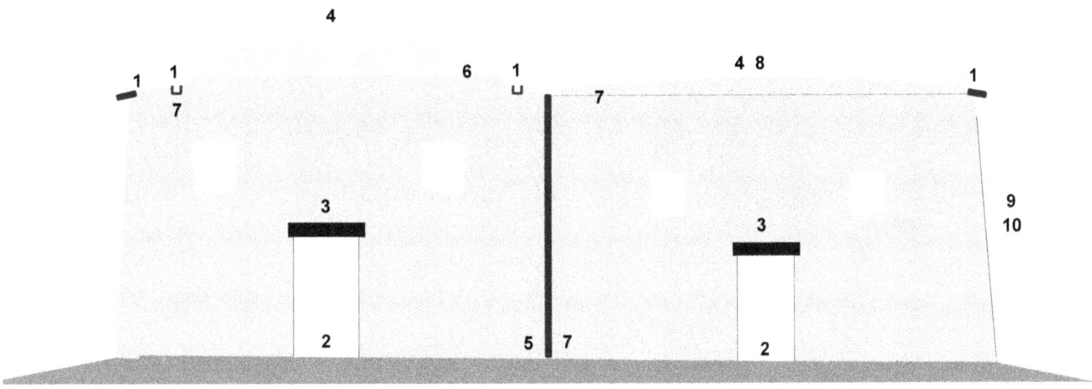

Sketch 1. Maintenance of earthen architecture.

1) *Wide drainage spouts of wood or ceramic are installed on rooftops and vertical surfaces. Regular maintenance checks, and where necessary repair are essential for these to remain functioning, and in instances where these are poorly maintained the subsequent damage caused may be excessive. Simple preventive measures such as checking gutters, and removing vegetation growth are important to prolong the life of an earth structure.*
2) *Regular (and daily) maintenance associated with the cleaning of areas more prone to gather dirt, such as cooking areas, ovens and thresholds - in these places daily maintenance might be through the application of a thin wash of earth plaster or earth and water.*
3) *Replastering of interior and exterior walls – the renewal and reworking of the surface, particularly in mudbrick structures where the exterior surface is coated in earthen plasters and renders, and for the interior of structures to keep them clean and functioning.*
4) *Roof maintenance of flat and domed and barrel roofs, through the use of earth plasters on top of the existing surface (which has been swept clear of debris).*
5) *More substantial repairs to wall bases through the cutting out of the damaged zone, and the insertion of replacement materials - depending on the area of the structure damaged this may again be coated in an earth plaster.*
6) *More substantial repair of roofs and wall tops through the cutting out of the damaged zone, and the insertion of replacement materials - depending on the area of the structure damaged this may again be coated in an earth plaster.*
7) *More substantial repair may routinely be required for stepped areas leading to roofs, areas of the external surface affected by dripping rainwater or to the base of undercut walls.*
8) *Re-compaction of the top layer of earthen roofs following precipitation, in some instances utilising roof rolling stones.*
9) *Maintenance is enabled as a result of access to the interior and exterior design at the type of construction.*
10) *Other patterns of maintenance and renewal may also develop cultural or socio-religious significance, and they are planned in association with annual festivities etc.*

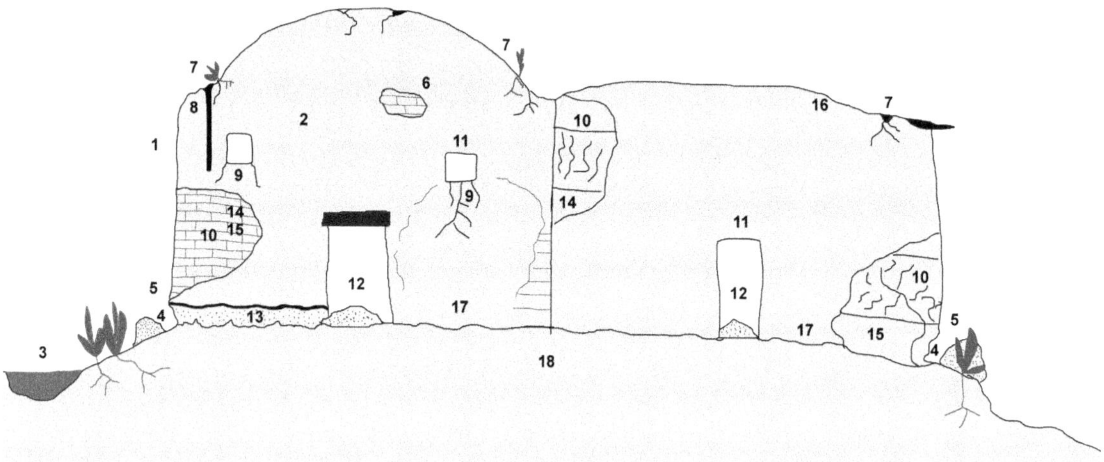

Sketch 2. Erosion factors effecting un-maintained earthen architecture.

1) *Causes of environmental deterioration are active, and earthen architecture is in a constant state of change.*
2) *Wind and wind blown sand cause differential surface erosion through abrasion, whilst precipitation washes fines out and obscures the surface of the structure.*
3) *Water trapped at the base of the occupation mound raises humidity in a zone at the base of the building*
4) *Capillary action takes place at the base of the building, and is exaggerated at the corners; the resulting piles of loose and spalled material raise higher the zone of potential damage.*
5) *Falling water creates a zone of damage at the base of the walls through splash back.*
6) *Extreme diurnal and annual temperature and climatic variation leads to the mechanical breakdown of the constituent parts of earthen material, resulting in erosion to the surface.*
7) *Lack of maintenance to the earthen roof allows the creation of cracks and water infiltration into the dome, increasing the risk of structural collapse at the junction between the dome and the vertical wall, in addition water run-off thins the top of the earth dome. In other areas lack of maintenance contributes to vegetation growth and the use of the eroding earthen material by insects and nesting birds.*
8) *Lack of maintenance to original drainage gullies results in erosion of the surface, first through the creation of runnels, and latterly through more substantial gullies.*
9) *Lack of maintenance and repair leads to construction imposed stresses through the structure; in particular rainwater gullies are quick to form at the base of voids, such as windows and doors.*
10) *Detachment of the earthen plaster/render from the surface of the structure through mechanical weathering, this exposes the wall core and earthen substrate to further damage from the effects of freeze/thaw, wind erosion and moisture. As a result of this exposure of the earthen substrate the mud brick or earthen mortar may deteriorate differentially, and insects and animals may utilise weaknesses in the earthen material.*
11) *Zones around voids, such as windows and doors, gradually erode to become thinner and wider, resulting in gradual enlargement through time.*
12) *As a result of loose collapsed materials, moisture becomes trapped in the interior, raising humidity and accelerating rates of erosion.*
13) *Capillary action leads to undercutting, and at the base of the structure surface detachment of the earthen plaster exposes fired bricks used in the original structure as a capillary break. The robbing of the fired bricks from the base removes the capillary barrier and increases the rate and height of capillary movement up the wall.*
14) *Stresses are imposed by the variation in the original material used for construction, such as the variation between mud bricks and mud mortar, and between the separate lifts in the rammed/ placed earth structure.*
15) *Detachment of the earthen plaster/render from the surface of the structure exposes the wall core and earthen substrate to further damage from the effects of freeze/thaw, wind erosion and moisture, through the exposure of the earthen substrate the rammed or placed earth lifts may deteriorate differentially, and insects and animals may utilise weaknesses in the earthen material.*
16) *Robbing and/or bio-deterioration of timbers from flat roofs leads to a rapid acceleration in erosion through the exposure of the wall tops to falling water and further erosion. Deterioration is accelerated through robbing and recycling of earthen material for incorporation in new construction or as fertiliser for agricultural fields.*
17) *Detachment of the earthen plaster/render from the base of the surface of the structure exposes the wall core and earthen substrate to more extensive damage from capillary action.*
18) *The processes of erosion and deposition lead to the accumulation of loose, more porous material both within the structure and in a zone on the immediate exterior. The moisture trapped in these erosion mounds moves the zone of evaporation up the remaining upstanding walls, and contributes to zones of undercutting higher up the structure. The more loose and moist materials also attract vegetation, altering humidity regimes, and further attracting insects and animals to the eroding structure.*

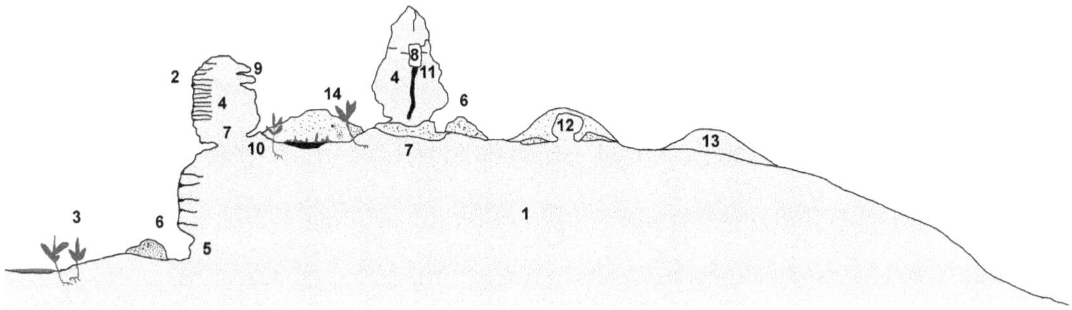

Sketch 3. Erosion factors effecting eroded but still extant earthen architecture.

1) *Causes of environmental deterioration are active, and the eroded earthen architecture is in a constant state of change.*
2) *Wind and wind blown sand cause differential erosion through abrasion damaging the upstanding earthen architecture, and by altering the shape and profile of the erosion mound through time.*
3) *Water trapped at the base of the occupation mound raises humidity in a zone at the base of the eroding building and mound.*
4) *Extreme diurnal and annual temperature and climatic variation leads to the mechanical breakdown of the constituent parts of earthen material, resulting in erosion to the surface.*
5) *Precipitation creates a zone of damage at the base of the walls through splash back.*
6) *Capillary action takes place at the base of the upstanding earthen structure and the resulting piles of loose and spalled material raise higher the zone of damage, and generate a deeper, and more extensive zone of damage.*
7) *Extensive damage to the base of the wall through undercutting alters the loading characteristics of the extant wall, making it more liable to dramatic collapse, under the influence of dynamic wind loads.*
8) *Birds nest in weak spots in earthen structure, such as original voids, and/or zones of differential erosion between mortar and mud bricks, the resulting phosphate rich surface deposits from bird excrement alter the chemical composition of the surface creating a zone with an increased propensity for erosion.*
9) *Insects such as hornets and wasps nest in earthen architecture, reworking the earthen material and re-depositing it in glass-like honeycomb structures.*
10) *Animals and reptiles excavate burrows and nests; the deposition of faeces alters the chemical and physical properties of the surface. In addition larger animals, such as camels, rub on exposed upstanding earthen walls and cause abrasion.*
11) *The original construction and design imposes stresses through the extant structure, in particular extensive run-off gullies formed at the base of the original windows channel run-off down the surface of the structure. Zones around voids, such as windows and doors, become thinner and wider, resulting in gradual enlargement through time.*
12) *Falling water washes fines out and obscures the surface of the structure, this results in the thinning and shortening of the walls. Surface run-off and wind erosion results in the movement of the eroded earthen material over the site and contributes to the exposure, covering up, and burial of earthen walls.*
13) *Continued surface erosion of extant walls eventually results in the formation of low mounds, associated with prevailing wind direction, and direction of surface run-off*
14) *Vegetation may add stability to the erosion material, and erosion mound, whilst other vegetation, such as tamarisk may contribute to changing moisture regimes on the surface, whilst the extensive root systems of desert plants can damage buried earthen material.*

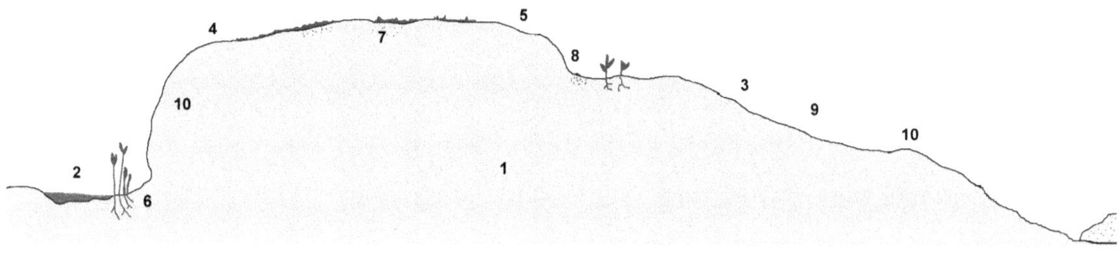

Sketch 4. Erosion factors effecting unexcavated archaeological sites.

1) *Causes of environmental deterioration are active, and the eroded archaeological site is in a constant state of change, altering the shape and degree of preservation of the unexcavated archaeological strata.*
2) *The proximity to stagnant water increases the overall porosity and looseness of the eroded earthen material, this creates an erosive matrix, which may be washed away as a result of surface run-off and wind erosion.*
3) *Wind and wind blown sand cause differential erosion, and through abrasion alter the shape and profile of the erosion mound through time.*
4) *Extreme diurnal and annual temperature and climatic variation leads to the mechanical breakdown of the constituent parts of earthen material, resulting in erosion to the surface.*
5) *Capillary action results in the deposition of salts at the surface or subsurface.*
6) *Vegetation growth is encouraged by proximity to water. Vegetation damages unexcavated earthen architecture as a result of root damage, and changes in microclimate association with water vapor and humidity on the surface.*
7) *Low lying grassy vegetation adds stability to gentle slopes and surfaces dependent on climate and local environment.*
8) *Animals, birds, insect and reptiles burrow into the softer eroded earthen material, adding chemical variation to the surface through waste deposits.*
9) *Gullying is caused by surface run-off across the loose erosive matrix down the eroded hill slope through hill wash.*
10) *'Tell-creep' occurs through the erosion on the wind buffeted side of the slope creating a zone which is poorly preserved, in contrast tell-creep alongside surface run-off re-deposits materials and creates a zone of the erosion mound which is better protected.*

Chapter 6: Earthen architecture in contemporary society

"the Persians are very carfull of their Terraces or Coverings of the Houses, as the chief Part, whereon depends their Preservation. Their Care about them, is to keep always the Rain-Spouts clear, at the bottom, and to sweep the Snow off the Terrace, when it falls very thick. 'Tis a Sport for the Mob to throw the Snow of the Houses, and they run up cheerfully to the House-top. The young men of the Ward go up into all the Terraces, one after another, and clear them in a Moment; and to encourage them the more to it the Musick waits on them all the time." (Sir John Chardin 1677, 263-264).

John Chardin describes the maintenance and repair activities associated with earthen architecture. The regular maintenance and less regular repair activities are essential to prolong the life of earth structures. The regulators of this life cycle are people and communities. People and communities play the key role in construction, and retain earth structures through vigilance, maintenance and repair activities. Disruption to the maintenance cycle interrupts the life of a structure, and initiates patterns abandonment and deterioration of earthen structures (see *Chapter 5)*.

The key aspects of maintenance are (1) a knowledge of when and how (through maintenance checks and the retention of skills), and (2) access to maintenance and repair materials (generally those from which the structure is built - earth, water and straw). Typical maintenance strategies include the replastering of interior and exterior walls, the repair and replastering of flat, domed and barrel roofs, and repairs to wall bases, alongside maintenance checks to thresholds and drainage. The appropriateness of these approaches for the conservation and management of earthen architecture in other contexts may not always be clear.

Given the properties of earth used as a building material, maintenance is required in order to moderate and remedy its active erosion characteristics. The types of erosion and the factors that cause deterioration occur in 'living' as well and 'abandoned' contexts. However, in 'living' contexts damaging effects are checked and managed through regular maintenance and corrective repair. *Chapter 5* identifies the active nature of earthen architecture, and discusses the patterns of erosion, loss and deterioration associated with earthen architecture once it has fallen out of use.

Corrective maintenance is required as a result of: the continued erosion of the surface through water run-off and the washing out of fines from the surface of the structure; erosion at the base of structures due to capillary action; erosion of the roof and wall tops; and erosion through wear of thresholds and floors, windows, drainage gullies, and stairwells. Some of the erosion most associated with earthen architecture is change to the appearance of the surface over time; a change which is rarely damaging by itself and is a natural characteristic of exposed earthen building materials. Nevertheless, the maintenance activities most usually associated with earthen architecture are the renewal and reworking of the surface, particularly for mudbrick structures where the exterior surface is coated in earthen plasters and renders, and within the interior of the structures themselves to keep them clean and functioning (Sketch 1).

The regularity with which maintenance and repair is carried out is determined by both environmental factors and stress (see *Chapter 5*), alongside social, cultural and economic factors. In addition, the nature of the structure, and the earth construction techniques from which it is formed, will determine the type and nature of maintenance and repair. For example, many structures comprising placed or rammed earth techniques, alongside those with decorative brickwork, or non-load bearing boundary walls, exist without exterior earthen plasters or renders.

Maintenance is a necessary activity for earth structures, however it is often seen as something automatic and minor - echoing the thoughts, "repair is a small matter not worthy of discussion." (Master builder Ramadān Rajab ba Shāmkhah to Damluji 1992, 395). The necessity for maintenance implies it is part of the world of unconscious, short-term habitual actions (Gosden 1994). A distinction can usefully be drawn between the daily-routine maintenance associated with cleanliness and custom; those activities associated with remedying the effects of erosion and weathering on earthen structures; and those pre-planned acts associated with more symbolic or social/cultural aspects of maintenance.

Daily routine maintenance

The habitual, daily actions concerned with keeping structures in good order might manifest themselves in the reworking of earthen floors through the daily acts of cleaning, brushing and wetting down, each wetting and drying episode re-compacting the floor. In other instances these regular acts of maintenance might be associated with cleaning of those areas more prone to gather dirt, such as cooking areas, ovens and thresholds. In these places daily maintenance might be through the application of a thin wash of earth plaster or earth and water. These kinds of daily activities can be interpreted from the microstratigaphy of the Neolithic structures at Çatalhöyük (Turkey), with a distinction seen between the multitude of plaster layers perhaps associated with daily activities, and thicker layers perhaps associated with seasonal and/or annual maintenance (*pers comm.* Wendy Matthews). There is also evidence of these sorts of daily activities in West Africa, with the application of a thin plaster or earth and water-wash over cooking areas (*pers comm.* S. Moriset). Similar daily, routine maintenance such as the whitening (or black leading) of doorsteps, windowsills, and hearths continued until the 20th century in the United Kingdom.

Maintenance and repair associated with weather and erosion

Practical acts of maintenance and repair occur in response to the gradual building up of the effects of weather and erosion on wall tops and roofs, wall bases, floors and wall surfaces. In most instances these acts of maintenance are associated with the reworking, reapplication, or recompacting of a surface. Simple preventative measures such as checking gutters, and removing vegetation growth are important checks that assist in prolonging the life of an earth structure (Walker *et al* 2005).

Most earthen buildings have mechanisms and adaptations built into the design and finish of the structure in order to combat environmental deterioration. For example, in wetter climates, foundations of stone or fired brick, and methods of dispersing water away from the main body, such as wide over-hanging roofs are common features (such as shallowly angled roofs in north-east Spain). In drier climates, characterised by a short, but intense rainy season, large drainage spouts of wood or ceramic are installed on rooftops (*canales* of the Southwest USA and *nāvdān* of Iran), and vertical surfaces (Fig. 28, 55-58). Regular maintenance checks, and where necessary, repair are essential for these to remain in good functioning order, and where they are poorly maintained the subsequent damage caused can be considerable, with substantial damage being caused by blocked water drains.

Environmental stresses determine the life span of the earthen surface finish, and subsequent patterns of maintenance and renewal, which may not be universal across the same settlement or structure. The regularity with which more substantial maintenance and repair occurs is dependent on local climate, customs and regularity of use (Horne 1994). For those earthen building materials that are covered with an earthen finish, the surface will require maintenance and renewal if it is to serve its protective function. Damage to the vertical surfaces of the structures is covered over with the re-application of earthen plaster. This maintenance involves the preparation of the earthen materials, and these materials are then applied over the surface, often with more substantial patches of damage receiving extra attention (Fig. 29, 30, 50-52). For example in north-eastern Iran exterior re-plastering utilising *kahgel* (earth plaster with straw), applied in a layer 1 to 3.5 cm thick has been observed as occurring every three years (Horne 1994 137, 141), and interior plastering with *gach-e khāk* (a lime plaster) occurs every 2-3 years (Horne 1994, 141) (which is distinctive to plastering associated with *Nowruz* see below). Within the study area the maintenance and renewal of the still-occupied earthen buildings is concerned with re-plastering and repair on a yearly or biannual basis. Roof maintenance involves the preparation of earth, with the addition of water and/or other additives. The prepared material is then applied over the existing surface (which has previously been swept clear of debris), and at this point areas of additional concern, such as drainage gullies can be addressed, again through the application of the wet earth mix, or with other repair materials, such as fired brick, the repaired zone subsequently being coated in an earth plaster as appropriate.

More substantial repair of wall bases, roofs and wall tops again occurs where needed, often involving the cutting out of the damaged zone, and the insertion of replacement materials. Depending on the area damaged this may again be coated in an earth plaster. More substantial repair may routinely be required for stepped areas leading to roofs, areas of the external surface affected by dripping rainwater or the base of undercut walls. Repair involves cutting out the damaged area, reworking the removed earthen material (if necessary with the addition of new earthen material), and infilling. If a repair is made to rammed or placed earth walls the infill material may be dried prior to insertion (as earth blocks) and the space between the new and old material filled with earthen mortar or plaster. After the repair has dried, and any further cracking filled, the patched area or entire wall surface can be re-plastered with an earthen surface finish.

Maintenance and repair associated with weather and erosion may also occur in response to immediate needs, and these types of maintenance may be associated with precipitation which damages wall tops and roofs, wall bases and wall surfaces. For example, earthen roofed structures in climates that may experience rain or snow, access to rooftops is essential for maintenance (Beazley and Haverson 1982). Roofs are particularly susceptible to damage, as through the absorption of water the clay component of the earthen material will expand and contract, losing its cohesiveness and compactness and start to erode. This damage is exacerbated when snow sits on the roof for a length of time, as this can also change the loading characteristics of the structure (see *Chapter 5*). The pattern of maintenance essential for these earthen roofs is to recompact the top layer of earth that will have absorbed the falling water. In these instances the task is to clear snow from the roof and to re-compact the surface, in some instances utilising roof rolling stones such as the Iranian *qaltabān* (Wulff 1966; Beazley and Haverson 1982).

A requirement of maintenance is access to the interior and exterior of the structure and as such, maintenance needs must be incorporated into the design of the earth structure. For example putlog holes, left over from the original building, are sometimes left topermit easier access, whilst for the tall, rammed earth structures of Morocco the holes left by formwork ties allow the easy introduction of new scaffolding for the purpose of maintenance and repair. Another method of providing access for maintenance is the insertion of reinforcing timbers as a permanent scaffold for maintenance needs, for example the mostly functional but partly decorative *toron* on the mudbrick mosques in West Africa.

Structures that remain only in occasional use may have different patterns of maintenance and repair as they do not have regular maintenance checks.

Symbolic, social and cultural aspects of maintenance

Patterns of maintenance and renewal may also develop socio-religious significance. For example, distinction can be made between the routine maintenance and repair of exterior walls, and the replastering of interior walls with *gel-e sabz* (green clay plaster) which occurs annually (by oral tradition) at the time of the Persian New Year – *Nowruz* - a time of rebirth and refurbishing (Horne 1994, 141).

To fail to maintain, or to let a building become damaged by the effects of water can be a controversial political act. An example is the contested history surrounding the maintenance, destruction and subsequent rebuilding of the Great Mosque at Djenne (Mali). One account of the building's history details the blocking of the gutters in the 19th century in order to let 'nature' destroy the structure (otherwise forbidden by Islamic law) (Bourgeois and Pelos 1996, 127-155).

The notion of maintenance as an important element of many religious or ritual practices has also changed during the 20th century, effecting the survival or otherwise of structures. For example, the Islamic system of *waqf* provided maintenance funds for edifices of a social or religious nature through trust funds raised through agricultural and commercial enterprises. This system supported the mosque it was set up to fund, the Maddrassah from which it gained employees, and the traditional houses and bazaar from which the income is generated (Hillenbrand 1994). The upkeep of both the institution and the income-producing buildings was the obligation of the trustees. Over the course of the 20th century the nature of *waqf* altered and evolved, and in some cases disappeared and dissolved, and in other countries was de-secularised and nationalised, as part of government ministries. The result is a change in the provision for maintenance for the mosque, Maddrassah, bazaar and domestic structures.

There is some evidence to suggest that the types and patterns of maintenance are also affected by gender roles. There is evidence to suggest a male role in manufacture and construction, with most apprentice schemes being male-only (for example Bedaux *et al* 2000), and a female role in decoration and final finishing (for example Rainer 1993).

Maintenance is sometimes seen as a female activity (as it is the decoration and final finishing that requires most maintenance), whilst other aspects of repair (such as for roofs that require more substantial work) may be male activities (for discussion of these roles at Merv see below). The gender roles assigned to labour and maintenance have a significant impact on the survival or otherwise of earthen architecture, as during the 20th century male members of society may have undertaken economic migration to urban areas, this influences the types and pattern of maintenance activities in traditional rural societies (see below).

In the course of the 20th century extra dimensions have intensified the relationship between maintenance, use and abandonment, such as the introduction of concrete at the expense of earthen surface finishes and the socio-economic change. The changing nature of the relationships between people and buildings is shown through alterations in the maintenance regimes appropriate for earthen architecture. In many locations around the world there has been a significant shift from the use of earthen plasters and renders to the use of harder, cement-based materials which are perceived to be longer lasting than the traditional materials they replace. This is problematic as the use of earthen plasters and renders on earthen structures allows the buildings to 'breathe' and to regulate moisture regimes naturally within the construction materials. In those instances where harder, cement-based materials are introduced they create a harder, impermeable barrier below which there is an increased rate of erosion and deterioration of the earth wall (see *Chapter 5, 9, 10 & 14*).

These relationships that influence use and maintenance are contextually dependent, but when they alter, and maintenance stops, and a building is abandoned or falls out of use, the earthen material may be rapidly reworked or may rapidly erode, leaving upstanding ruins, or through time archaeological deposits of earthen architecture (see *Chapter 5*).

Living skills – knowledge transfer and apprenticeship

The skills associated with the construction, maintenance and use of earthen architecture are transferred between generations in informal contexts, in which everybody in society is expected to be able to provide for their own shelter (Oliver 2003), or in more formal apprenticeship contexts through guilds and trade organisations (*op cit*; Marchand 2001; Argumendo 1981). Oliver (2003) observes differences in the status of construction dependent on the level of formalisation of the industry. For example, where building is held in low-status by society everybody is expected to contribute to construction, this in contrast to more formalised guilds and associations, in which skills are passed on through apprenticeship. The skills themselves are learned through experience, and also transmitted through memory training, as songs, proverbs and repetitions (*op cit* 81-83).

Contemporary earth use at Merv

The current setting of the archaeological park is characterised by small-scale agriculture and former collectivised village communities (*kolkhoz*) established from the 1930s. For the purpose of gathering this material the most important of these communities is the Ancient Merv *kolkhoz*, which is one of oldest *kolkhoz* in the Merv Oasis (*pers comm.* Rejeb Dzaparov). This *kolkhoz* is characterised structurally by the use of a variety of different materials and techniques for the construction of single-storey rectangular structures to create family compounds. There are no legal requirements for construction to follow, although social, cultural and economic factors result in a general, overall homogeneity.

Construction materials

Contemporary construction utilises a variety of different materials, and the modern buildings often use a combination of building materials, such as: fired brick *Galtak/Bishen kerpic,* of a standard size c. 240mm x 124 mm x 60mm introduced from about the 1950s (*pers comm.* Rejeb Dzaparov); cement (introduced through the 20th century imported from Iran, Kazakhstan or Uzbekistan, and now manufactured outside the Turkmen capital Ashkabad); and earth in construction. There is contemporary evidence for a number of different earth building traditions used for both load bearing and nonload bearing construction. These include mudbrick (*gala kerpic*), placed earth and rammed earth (*paksha*), earthen mortars, plasters and renders *(suwoq/suwool gelina)*.

The modern *gala kerpic* are small (240mm x 124 mm x 60mm), formed in the same moulds (*galep*) as modern fired bricks, and referred to as 'Russian bricks' (however the *gala kerpic* may not necessarily be standardised as everybody uses and makes them*)*. Sometimes there may be alternate *paksha* and *gala kerpic* construction. Today *pakhsa* construction is in decline, and in the Merv village individuals plan to replace the older eroding *pakhsa* buildings with new structures which combine *gala kerpic* and fired bricks (Fig. 43).

In the village locations surrounding the archaeological park most new construction uses concrete as a building platform, and a majority of the mudbrick buildings now have fired brick facades (Fig. 42-45). When fired and mudbricks are combined in the same structure, every fifth brick course is connected, and/or the face of the structure is built up in fired bricks. The use of *gala kerpic* is a cheap way of adding overall height to a structure, using the fired bricks for the foundation and bottom courses and then *gala kerpic* for the rest of the wall, whilst *gala kerpic* are used for interior walls. In this way a majority of the buildings in the Merv oasis appear from the exterior to comprise solely of fired brick construction. The utilisation of fired brick in combination with *gala kerpic* is attested to utilising both the thermal characteristics of the *gala kerpic* alongside the perceived benefits of reduced maintenance in the use of fired bricks (*pers comm.* Rejeb Dzaparov) (although interestingly it is the comfort and 'livability' rather than the economics of thermal regulation that are key, as both electricity, and gas (since 1992) are free (*op cit*)).

Construction practice

Earth buildings are built from about May (the end of the rainy season) and generally finished in November (the start of the rainy season) (although *pakhsa* tends to be carried out only in the high summer (*pers comm.* Akmohammed Annaev). The main masonry component of the building is normally complete in 1 month, after which the interior, exterior, flooring and roof is finished.

Modern earth construction uses the silts that accumulate from the cleaning of irrigation canals and ditches. Preference is made for locations with 'sweet soil' (earth free of salts, agri-chemical residues and without too much sand). Earth is mixed and prepared in a '*handek*' either on or off site, depending on proximity to water and the nature and type of earth available on site. If they are off site, they are often near water channels or water supplies. *Gala kerpic* tend to be made off site, utilising locations adjacent to modern irrigation canals, which have quantities of up-cast silts available; in contrast earth is transported and *paksha* is mixed on site ready for construction. In addition material from older eroding buildings is sought out and recycled for mudbrick manufacture. The re-use of the older material is attributed to the easier working of the already 'cooked' older material, additionally there are benefits in using material that has already been worked and in which materials such as straw have been added. In other instances the material from older eroding buildings is used for the construction of the building platforms (Cooke 2004).

Today the most common additive to the basic earth and water mix is finely chopped wheat straw (*saman*) for use in *gala kerpic* manufacture and for earth plasters (*paksha* generally has no plant materials added). The straw is chopped into very small pieces 0.5-1.0cm length and 0.1-0.2 mm width. The *saman* used in earthen plasters is a byproduct of flour/bread making and is ideally left to rot for 12 months prior to use, and after mixing is left for 1-2 days, normally until it smells (this is locally attested to preventing the straw from sticking to any of the plastering tools) (*pers comm.* Dowran Durdiev). The preparation and mixing of the basic earth mix generally take 3 days, with the soil, water and *saman* mixed and left on day 1 and 2, ready for use on the third day, when the mixture is tested to ensure it is suitable by stretching it in a rectangle, which two people pull: if it breaks in the middle the mixture needs more work (*op cit*). Using this basic earth mix, construction can occur in a variety of different forms, either as mudbricks, *paksha* or plasters (although the top layer of the earth plasters contain much finer lengths of straw than those layers underneath).

The process of construction involves the creation of a platform that consists of earth (*komkowy*), or concrete. This platform is constructed several months prior to commencing the rest of construction in order for it to dry and settle. Foundations are dug and built through the platform, and on top of this foundation asphalt paper or liquid bitumen is added as a capillary barrier. *Gala kerpic* are laid with a mud mortar, and depending on the position in the structure (and whether or not fired bricks are utilised) the exterior is coated in a mudplaster, whilst the interior is coated in a thin white wash. *Pakhsa* walls are built up in layers of descending height (1st = 90 – 100 cm, 2nd = 70-80cm, 3rd = 60-70 cm, 4th = 50 – 60 cm.), after 2-3 days of drying the wall is levelled, following another 7-10 days for each layer to dry prior to proceeding with the next. Wood is used in between the *paksha* lifts. This is often tamarisk, but mulberry is preferred, as it is very strong and not attacked by insects (mulberry wood is also used for reinforcement when concrete is used for construction) (*pers comm.* Akmohammed Annaev). After 1 year the *paksha* wall is thought to be dry, and the cracks that formed in the drying out process are then filled with earth plaster, after which the entire wall may be coated in a thin white-wash.

The older buildings in the villages have large flat roofs constructed with large wooden poles, and a reed lattice, upon which layers of mud plaster are placed. The roofs have a 30cm slope on both sides in order to facilitate water run-off and drainage pipes are also installed. These flat roofs continue to be used for storage of winter fuel and fodder, such as camel thorn. Whilst the flat roofs are most often associated with those structures comprising *gala kerpic* or *paksha* construction, a number of buildings with a fired brick exterior also have flat earthen roofs. Today, depending on the wishes of the owner, a roofing superstructure could be erected over the earlier roofs, and more commonly, new construction utilises pitched (inverted v-shaped) roofs. These roofs are constructed out of roofing timbers, and covered with corrugated metal (and in some instances asbestos) sheeting. These inverted pitched roofs are expensive, and are not very well suited to the local climate, as they do not moderate the diurnal and annual temperature fluctuations, however they are often utilised as a result of socio-economics perceptions and change (*pers comm.* Myrat Kurbansakhatov).

For storage and ancillary structures, and for non-load bearing walls comprising *gala kerpic* and *paksha* the exterior and interior surfaces may not be coated in earthen plasters and/or white wash. Non-load bearing walls often have a capping of loose brushwood and earth plaster.

Construction organisation

Modern construction in Turkmen villages involves the employment of a master builder, alongside both skilled craftsmen, and family members. There are perhaps 50 different people in the local *kolkhoz* with the skills required to build in earth, including mason, helpers and workers (*pers comm.* Rejeb Dzaparov). Most male family members have been involved in the construction or maintenance of dwellings, outbuildings and additional structures in the family compound. There is very little formal training amongst the owner builders on techniques, although there is a great deal of knowledge about sourcing of suitable earth and simple methods of testing the suitability of earth used in construction (such as pulling the earth, length of straw and degree of mixing).

Women are often, but not always attributed with internal decoration (painting and applying white washes and wallpaper), and the construction of earthen bread ovens (*tamdyrs*). The division in labour associated with traditional Turkmen family life, reflecting the traditional saying: "The world is a man's house, while the house is a woman's world." (Blackwell 2001, 149). For example, the repair and maintenance of the earth roof on the Mausoleum of Ibn Zeid involved the masons employed by the archaeological park discussing methodologies and earth plaster mixes with their elderly female relatives, who had more experience of building maintenance (*pers comm.* Sebastien Moriset; Tim Williams) (the assumption was that only men are involved in the maintenance, as only men are involved in construction (*pers comm.* Rejeb Dzaparov).

Maintenance

For *gala kerpic* buildings the ideal is to maintain walls and roofs with the reapplication of mud plaster every year, and for *paksha* buildings through the reapplication of the thin whitewash, alongside the corrective repair of areas of more substantial damage. However the regularity with which this occurs depends on economic factors and the availability of labour. The maximum periods without maintenance for walls with no capping is generally every year, otherwise depending on orientation, an interval of up to 10-20 years for the reapplication of earth plasters and 4-5 years for whitewashing.

The requirement for maintenance of the earthen surface finishes (plaster and render) depends on the type of earth construction method used, with *paksha* structures often without an exterior render of whitewash, and function of structure, and structures that are not lived in (even mudbrick structures), such as cattle and storage areas may not have exterior or interior plaster.

Although the timing and regularity of maintenance is not directly associated with ritual or religious activities, there is the possibility that in the past some of the maintenance activities were associated with particular events, such as *Nowruz* (see Blackwell 2001).

Other construction

A majority of houses have an earth *tamdyr* (bread oven) (Fig 41) The *tamdyr* is constructed first with a ring of fired bricks, with mud plaster on either side, then prepared earth is placed by hand to build the body of the oven, and the whole then coated in earth plaster. There is anecdotal evidence to suggest that women construct and maintain the earth ovens, whilst men dismantle them to avoid bad luck The fuel now used in the *tamdyrs* is waste left-over from the growing of cotton (as this burns slowly making it ideal for bread baking).

A very different use of earth is represented in Turkmen funerary rites, where burial occurs beneath a low earthen mound (Blackwell 2001; Fig. 46 & 47). The body is carried to the burial place on a ladder, and the ladder is left jutting out of the mound. In some places these funerary mounds are also coated in an earthen plaster, which is reapplied annually (*pers comm.* Gaigysyz Joraev). Showing similar patterns associated with the changing nature of domestic earthen architecture this earth plaster is now being replaced with a thin cement capping (Fig. 46).

Symbolism of earth

Within traditional Turkmen society there is a rich symbolism associated with earth structures. Blackwell's

(2001) folklore study of spontaneous women's songs analyses them from a contemporary cultural context. Within these songs there are references to buildings and the home, and within these songs the use of metaphors associated with earthen architecture is significant. For example, within the *Laeleler* (girl's songs which reveal their thoughts and concerns):

Long long walls of clay.
My brother binds his feet with cloth.
Until my brother dismount from his horse,
My heart bursts with impatience.
(Blackwell 2001, 99)

Blackwell interprets this song as the long earth wall, which in traditional Turkmen society kept a girl inside, whilst allowing her brother to be free to travel on long journeys.

Within *Huewdueler* (mothers lullabies):

My little brother is like an Arab,
Mounting a horse suits him.
From the clay towers in the yard,
His betrothed stands watching him.
(Blackwell 2001, 151)

Blackwell interprets this lullaby as about a wedding, and clay towers symbolise prosperity, so the lullaby is concerned with the becoming a good horseman, and growing to take a bride from a good family.

Within *Agylar* (lamentation or weeping songs):

The mound of my child lies ahead,
I can see part of it.
Other troubles have remedies,
But the mound of a child has no cure
(Blackwell 2001, 165)

Blackwell interprets this lamentation as a play on the '*dag*' - the mound (and also brand on cattle), so the funeral mound is flattened over the course of time, but the wound caused by the death of a child remains forever.

Within these traditional songs the metaphors associated with earthen architecture indicate that the earth walls can both protect and entrap, that earth structure can be linked to prosperity, but that like the earth burial mounds they may erode away.

The maintenance of structures has always been recognised as essential for the retention and conservation of the archaeological and historical environment. Maintenance is concerned with (1) maintenance of use, and (2) regular maintenance activities.

Maintenance of occupied structures is recommended within current conservation charters, thus the 1931 Athens Charter states:
"The conference recommends that the occupation of buildings, which ensures the continuity of their life, should be maintained but that they should be used for a purpose which respects their historic or artistic character." (Principle 1).

This emphasises the apparent appropriateness of maintaining occupation for structures (as occupation implies maintenance checks and regimes). In addition maintenance has been seen as appropriate for retaining structures, and the comparison and contrast between maintenance and restoration was used within the 19th century conservation debate. John Ruskin polemically argued against restoration, and he was in favour of maintenance:

"The principle of modern times, (...) is to neglect buildings first, and restore them afterwards. Take proper care of your monuments, and you will not need to restore them. A few sheets of lead put in time upon the roof, a few dead leaves and sticks swept in time out of a water-course, will save both roof and walls from ruin. Watch an old building with an anxious care; guard it as best you may, and at *any* cost, from every influence of dilapidation. Count its stones as you would jewels of a crown; set watches about it as if at the gates of a besieged city; bind it together with iron where it loosens; stay it with timber where it declines; do not care about the unsightliness of the aid: better a crutch than a lost limb; and do this tenderly, and reverently, and continually, and many a generation will still be born and pass away beneath its shadow." (XIX The Lamp of Memory Ruskin 1849).

Echoing these thoughts, William Morris, comments:

"It is for all these buildings, therefore, of all times and styles, that we plead, and call upon those who have to deal with them, to put Protection in the place of Restoration, to stave off decay by daily care, to prop a perilous wall or mend a leaky roof by such means as are obviously meant for support or covering, and show no pretence of other art, and otherwise to resist all tampering with either the fabric or ornament of the building as it stands; if it has become inconvenient for its present use, to raise another building rather than alter or enlarge the old one; in fine to treat our ancient buildings as monuments of a bygone art, created by bygone manners, that modern art cannot meddle with without destroying." (SPAB Manifesto 1877).

Therefore maintenance of buildings in use, using appropriate materials and techniques by skilled craftspeople, is in accordance with conservation theory. However, it is when maintenance regimes and patterns change (through discontinuation or abandonment) that problems are posed. For example, difficulties are posed when the nature of the maintenance regime alters (such as through the development of modern materials and techniques). Within the study area this problem is typified by the use of harder cement-based materials for surface renders. More generally the construction and maintenance skills associated with earthen architecture have altered, affected by outside influence and social, cultural and economic change. For example, agricultural communities may suffer from the depopulation of the economic migration of the male members of society, as a result traditional materials and techniques may be replaced with harder cement renders perceived to be more long lasting and/or displaying economic prestige. In these instances the alteration of maintenance materials and techniques in living contexts is problematic and poses problems in relation to current conservation theory. These factors will be discussed below highlighting how these altered maintenance regimes challenge, and are challenged by, current conservation theory.

PART 2

Chapter 7: Conservation, management and heritage studies.

This chapter discusses the emergence and development of conservation and management approaches to the historic and archaeological environment. The first part of this chapter discusses the development and characteristics of the international/western approach to conservation. The second part addresses the development and characteristics of current conservation theory. This chapter concludes with a discussion of current conservation theory and contemporary society. Throughout the chapter the key chronological developments are outlined in Table 3, this allows the chapter to discuss and explore the characteristics of conservation theory and ethics.

Whilst it is slightly cumbersome to retain a chronological split between these approaches to conservation theory (and in many ways repeats knowledge that is already well established as the basis of the conservation discipline), it is important to understand the development of the international/western approach to conservation as contemporary approaches to conservation philosophy are developed from this framework.

The development of the international/western approach to conservation

Our 'past' is created, studied and retained through conservation. The archaeological record itself has been used to suggest that the awareness of the past is a universal phenomenon and one that is evident since prehistory (Bradley 2002; Gosden 1994). Our different 'pasts' are as diverse as the numerous methods and approaches to its conservation. The development of one such approach to conservation - the international/western approach is well documented (Jokilehto 1999). This approach commences during the Renaissance in the 15th century. At this time Western European culture believed in a classical inheritance and values were ascribed to historic monuments as sources of inspiration, affirmation and as testament to humanity and civilisation. As such a requirement grew for objects or monuments to be kept for future generations to inherit. One of the earliest examples of conservation dates from the papal re-occupation of Rome in 1420. At this time attempts were made to limit the destruction of the archaeological and historic monuments threatened by quarrying for lime production (Choay 2001).

Much of the literature concerned with the development of conservation approaches places emphasis on the emergence of modern society through the agricultural and subsequent industrial revolution. These resulted in people becoming disconnected from their past, and subsequently concerning themselves with its study, revival and retention (Wetherall 1998; Brooks 1998). Both scientific and sentimental interests in the past were concerned that aspects of the past should be kept for future generations (Jokilehto 1999; Macaulay 1953). The combination of learning and aestheticism resulted in tensions as to what we do with and how we treat the things from the past (Denslagen 1994; Choay 2001). An example of these tensions is the concern with aspects such as anti-restoration and authenticity (in contrast to 'invention', Hobsbawn 1983). Key events in the development of approaches to conservation are described in Table 3.

Characteristics of the international/western approach to conservation

The characteristics of the international/western approach to the archaeological and historic environment can be categorised as:

Anti-Restoration

The anti-restoration approach grew out of concern with how, and in what ways the past might be kept. In the United Kingdom the approach emerged in the later half of the 19th century from a growing awareness of the scale and rapid increase of restoration carried out on later medieval ecclesiastical buildings (Miele 1998), alongside the growing multitude of 'invented' interpretations of the past in restoration work (characterised by Sir Giles Gilbert Scott, William Burges and, on the continent, by Eugéne-Emmanuel Viollet-le-Duc) (Thompson 1981; Denslagen 1994).

The movement was already established when it found its most vocal supporters in John Ruskin and William Morris (Chitty 1987). In this context John Ruskin established architecture as a moral, Christian presence in society with the publication of his *Seven Lamps of Architecture* (1849). Ruskin found inspiration in the work of Robert Willis who created and popularised the study of buildings archaeology and the idea of building narrative (Thompson 1981). Ruskin argued for the 'truth' in buildings, and the notion of a building's information and documentary value as being the significant aspect of the buildings life. As both restoration and reconstruction would impact the 'truth' and 'life' of the building Ruskin argued both were to be violently opposed:

"Neither by the public, nor by those who have the care of public monuments, is the true meaning of the word *restoration* understood. It means the most total destruction which a building can suffer: a destruction out of which no remnants can be gathered: a destruction accompanied with false description of the thing destroyed. Do not let us deceive ourselves in this important matter; it is impossible, as impossible as to raise the dead, to restore anything that has ever been great or beautiful in architecture." (Lamp of Memory XVIII).

William Morris masterminded, and rallied his friends, to support the foundation of the Society for the Protection of Ancient Buildings (SPAB), and its manifesto appeared in 1877. The manifesto states that each building has a life, formed of the changes that have occurred in the style appropriate to the period of alteration. Morris saw both destruction and restoration as having the same disastrous effect of stripping the life out of the building, resulting in

	KEY EVENT	EFFECT	GENERAL TREND
1500 -			Agricultural and Industrial Revolutions result in awareness, study and use of the past.
	1849 - John Ruskin *Seven Lamps of Architecture.*	Establishes architecture as a moral, Christian presence in society.	Growing awareness of the scale and rapid increase in restoration and reconstruction.
1850	1876 - William Morris voices fury over restoration of Burford Church, Oxfordshire.	Influences decision to establish the Society for the Protection of Ancient Buildings (SPAB).	
	1877 - SPAB Manifesto.	Results in greater public debate about conservation approaches	
	1879 - SPAB public protest against the restoration of St Mark's, Venice.	Results in greater international debate about conservation approaches.	Growing awareness of international threats.
	1882 - UK Ancient Monuments Act	Establishes the legal basis for philosophy of care and treatment of ruins in the UK.	The application of anti-restoration approach to archaeological sites and historic buildings
1900	1904 - The International Congress of Architects 4th meeting in Madrid	Sets out an anti-restoration framework alongside concern for the historical, technical and aesthetic values	
	1931 - International Congress of Architects and Technicians of Historic Monuments meet in Athens.	Recommends the use of new and scientific materials, problems of environmental deterioration, stresses the importance of the historic fabric, *anastylosis,* and the use of distinguishable materials.	
1950	1956 - UNESCO Recommendation on International Principles Applicable to Archaeological Excavation	Suggests common international principles for the provision of excavation, restoration and conservation.	Increasing emphasis on internationalism and the idea of a 'common' cultural heritage
	1964 - Venice Charter	Embraces technology and science, modification and change of use, *in situ* preservation, restoration and conservation, with the specific recommendations such as the use of *anastylosis*.	
	1972 - UNESCO Convention Concerning the Protection of the World Cultural and Natural Heritage	Establishes the World Heritage list and introduces notions of universal value and authenticity.	
	1990 - ICOMOS Charter for the Protection and Management of the Archaeological Heritage	Emphasises *ex situ* reconstruction for experimentation and interpretation	
	1994 Nara Document on Authenticity	Introduces a broader concept of authenticity, which includes form and design, materials and substance, use and function, traditions and techniques, location and setting, and spirit and feeling.	Questioning of notion of 'universal' value. Debate on Authenticity.
	1999 Australian ICOMOS Burra Charter (Revised)	Introduces value-based management planning; States "The aim of conservation is to retain the cultural significance of a place."	Increasing knowledge of exclusion, reflected as a need for participation.
2000	Various International, Regional and National statements for conservation	All emphasise values, heritage and participation.	Express the importance of cultural significance, value and authenticity on international and national approaches to heritage.

Table 3. Key events in the development of conservation approaches.

the loss of the buildings instructive, informative and associative value:

"those who make the changes wrought in our day under the name of Restoration, while professing to bring back a building to the best time of its history, have no guide but each his own individual whim to point out to them what is admirable and what contemptible; while the very nature of their task compels them to destroy something and to supply the gap by imagining what the earlier builders should or might have done. Moreover, in the course of this double process of destruction and addition, the whole surface of the building is necessarily tampered with; so that the appearance of antiquity is taken away from such old parts of the fabric as are left, and there is no laying to rest in the spectator the suspicion of what may have been lost; and in short, a feeble and lifeless forgery is the final result of all the wasted labour." (SPAB Manifesto).

Restoration was categorised by the SPAB as a "strange and fatal idea" (1877). The same anti-restoration ideas are reflected in the body of literature concerned with approaches to the archaeological and historic environment from a number of Western European countries (for example Riegl 1903, and other writings collated in Stanley Price *et al* 1996). It is also significant these anti-restoration ideas developed in the late 19th and early 20th century at the height of imperialism, as a result the anti-restoration approaches to the archaeological and historic environment were exported, by either colonial powers, or after independence, by retaining the legislation related to the historic environment (Ito 1996).

International conservation recommendations are characterised by retaining the anti-restoration emphasis. For example, the 1904 International Congress of Architects expressed an anti-restoration framework, highlighting concern for the historical, technical and aesthetic values as the important elements to be conserved. Similarly the 1931 International Congress of Architects and Technicians of Historic Monuments, recommended approaches within an anti-restoration framework, emphasising a desire to prevent total restoration and to retain the historical and aesthetic values, whilst limiting the negative impacts on its narrative (Burman 2003). This approach to conservation is characterised by a reluctance to rebuild or reconstruct missing elements of upstanding or ruined structures, alongside a reluctance to rebuild *in situ* on archaeological sites.

The most significant statement of international principles guiding approaches to conservation is the 1964 Venice Charter. This charter declares the purpose of restoration is to reveal the aesthetic and historic value of the monument. The charter states restoration, "must stop at the point where conjecture begins, and in this case moreover any extra work which is indispensable must be distinct from the architectural composition and must bear a contemporary stamp" (Article 9), whilst, "replacements of missing parts must integrate harmoniously with the whole, but at the same time must be distinguishable from the original so that restoration does not falsify the artistic or historic evidence" (Article 12). In this context the charter states new work must be in a contemporary style and be distinguished from the old so as to avoid the danger of falsifying the monument (Articles 9 and 12). The 1972 World Heritage Convention explores the same conservation philosophy, for example stating reconstruction is only acceptable if it is carried out on the basis of complete and detailed documentation on the original and to no extent on conjecture.

In many respects the approaches advocated in the international conservation charters reiterate the Ruskinian message that the protection of the tangible historic and archaeological fabric is a universal moral obligation and collective responsibility that should be carried out to retain the evidence of the past without altering it through restoration. This anti-restoration philosophy underpins many of the characteristic international/western approaches to conservation discussed below.

Archaeological and historic monuments

The international/western approach recommends different conservation methods for archaeological and historic fabric in different contexts (either in use or disuse). The 1904 International Congress of Architects divided the material from the past into two classes, with each class having different approaches to its conservation: 'dead monuments', were to be preserved and propped up by strengthening to retain their historical and technical value; and 'living monuments', were to be restored in their original style to retain their use, utility and aesthetic values (Recommendations of Subject II "The Preservation and Restoration of Architectural Monuments"). As such the international/western approach is primarily concerned with archaeological or historical structures that are valued for historic, visual or rarity reasons.

The archaeological and historic fabric may have gained new values through time, for example, Riegl classified the remains of the past into two categories, "deliberate" monuments and, "artistic and historical monuments" (1903, 1928). Similarly, structures have been classified as those that are built as monuments, with a purpose that is concerned with the act of memory or commemoration (such as war memorials and mausolea); and other sorts of structures, for Bradley the monuments with the "double lives" (2002, 82). These "double-lived" monuments are formed through time acquiring importance and value, signifying the past, but with very different values to those associated with the original use (which in many cases does not need to be known - as in the case of prehistoric henge monuments).

Iimportance of the archaeological or historic fabric

The international/western approach to conservation places great importance on the archaeological or historic fabric to ensure it is passed on to future generations, as a universal inheritance. The archaeological or historic fabric is thus seen as the most important aspect of the structure (Ruskin's 'living witness' or building narrative) - as such interference should be minimised. By placing importance on the archaeological or historic fabric the international/western approaches to conservation echo Ruskin's demand of monuments that "*We have no right whatever to touch them.* They are not ours. [Authors emphasis]" (1849, 197).

The importance of the historic fabric is re-iterated in conservation charters, such as the 1931 Athens Charter. This resulted in the development of particular conservation approaches such as stabilisation; a minimum approach; the freezing of the material evidence of the past in time; retaining the look of age (patina); retaining the visibility of interventions; ensuring reversibility; and the use of *anastylosis* (see below).

By trying to 'freeze' the archaeological and historic fabric in an 'as found' condition, these conservation approaches placed value on the archaeological and historic fabric, without impacting its narrative function, and emphasised the age value. For example, Riegl (1903) assigned age value as a visual component that is reliant on the look of age of the archaeological or historic fabric. Riegl commented:

"age value is revealed in the imperfection, a lack of completeness, a tendency to dissolve shape and color, [sic] characteristics that are in complete contrast with those of modern, i.e., newly created works." (1903, 1928, trans 1996, 73).

For Riegl age was an inclusive value as it is possible to know a building is 'old' without understanding other aspects of the structure. Age also appealed to the emotive and mystical senses, with age fulfilling a desire for atmosphere (as William Morris had already emphasised). The emphasis on age value is suggested through the importance placed on retaining 'patina' on the archaeological or historic fabric.

One aspect of the anti-restoration approach which suggests a more fluid approach to the 'conserve as found' doctrine is the use of reconstruction on archaeological sites. In this context the 1964 Venice Charter stated the purpose of archaeological excavation as one of information gain, dissemination and display, whilst article 7 of the 1990 ICOMOS Charter for the Protection and Management of the Archaeological Heritage emphasised the role of *ex situ* reconstruction for experimentation and interpretation.

Visibility and Reversibility

The 1931 Athens Charter advocated the narrative function of the historic fabric through the use of distinguishable materials in conservation work. This was encouraged in order to reduce the visual impact and to aid appreciation of the structures narrative. International/western approaches to conservation are similary driven by attempts to ensure reversibility. This emphasis is concerned that any modern intervention can be taken away and the condition of the archaeological and historic fabric can be 'returned' to. The philosophy of reversibility (like visibility) is influenced by the value placed on the visual narrative of the archaeological and historic fabric. The iconic example of conservation work that places emphasis on visibility and reversibility is the reconstruction of the Roman Forum (Italy) in the 19th century; here the architect's produced new elements in simplified shapes and distinguishable materials (Schmidt 1997, 41-42).

Authenticity

From the early 20th century the concept of authenticity grew to have great significance in international/western conservation approaches. This was heavily influenced by the reconstruction work carried out on classical Greek sites (Dimacopoulos 1985). The concept was established within conservation theory at an international scale, through the Athens Charter (1931), Venice Charter (1964) and World Heritage Convention (1972). For example, for the inclusion of sites on the UNESCO World Heritage list they must pass the 'test' of authenticity through the design, material, workmanship, distinctive character and components of a monument, building or site.

In these charters authenticity expresses itself in a notion of the genuine, the real and the opposite of fake. Authenticity implies having authority or coming from the author, it is often seen as grouped concepts; ancient and original in opposition to modern and reproduction; truthfulness in opposition to falsity. The international/western concept of authenticity is implied by patina, incompleteness, traces of wear and use, and uncontaminated contexts.

Anastylosis

The concept of *anastylosis* - the reinstatement of the fallen original fragments - was established within international/western conservation approaches in the 20th century. Although *anastylosis* can be documented on many sites, the re-erection of the columns on the northern side of the Parthenon (Greece) between 1922 and 1930 provides a significant example, and one that led to the formalisation of the approach within the 1931 Athens and 1964 Venice Charters (Schmidt 1997, 43-44). The 1964 Venice Charter states:

"All reconstruction work should however be ruled out "*a priori*." Only *anastylosis* can be permitted. The material used for integration should always be recognizable and its use should be the least that will ensure the conservation of a monument and the reinstatement of its form" (Article 15).

The use of *anastylosis* (as with visibility and reversibility) emphasises the importance placed on archaeological or historic fabric and the retention of the visual narrative for a site or structure.

New materials and a role for science and industry

International/western approaches to conservation often foresee a role for new materials, and input from science and technology for the design, type and nature of the materials and techniques utilised for conservation interventions. The 1931 Athens and 1964 Venice Charters are underpinned by a modernist approach and significant in proposing the use of modern materials and techniques in conservation:

"where traditional techniques prove inadequate, the consolidation of a monument can be achieved by the use of any modern technique for conservation and construction, the efficacy of which has been shown by scientific data and proved by experience" (Article 10).

There are numerous examples of using new materials, for example the so-called *anastylosis* carried out on The Library of Celsus in Ephesus (Turkey) which made use of reinforced concrete (Schmidt 1997, 46; Demas 1997, 140). The emphasis on the discovery of new conservation materials is a result of the search for long-term solutions for different materials, and the difficult demands placed by the (sometimes unrealistic) combination of international/western conservation theories and aspirations, such as minimum intervention, reversibility, and visibility.

International importance

From the outset international/western approaches to conservation have been concerned with international conservation efforts. In 1879 SPAB initiated a public protest against the restoration of St Mark's, Venice (Denslagen 1994). These international efforts were driven by a notion that as citizens of the world the past belongs to us all, and is valued by us all, and is to be inherited and valued by all future generations. This sense of universalism is reflected with the conservation charters, for example the 1931 Athens Charter highlighted international concern for conservation noting the universality of approaches and the nature of international co-operation and collaboration, ascribing the League of Nations as the "wardens of civilisation" (Article VII).

In the later half of the 20th century internationalism is expressed in the creation of institutions with global interests, such as the United Nations; and cultural heritage organisations such as UNESCO, ICOMOS, and ICCROM. The international documentation emphasises universality and common cultural connections, seeing heritage as a political message of unification. For example the preamble to the 1956 UNESCO Recommendation on International Principles Applicable to Archaeological Excavation states: "the feelings aroused by the contemplation and study of works of the past do much to foster mutual understanding between nations." Similarly the 1964 Venice Charter states:

"Imbued with a message from the past, the historic monuments of generations of people remain to the present day as living witnesses of their age-old traditions. People are becoming more and more conscious of the unity of human values and regard ancient monuments as a common heritage. The common responsibility to safeguard them for future generations is recognized. It is our duty to hand them on in the full richness of their authenticity." (Preamble to the Venice Charter 1964).

The international and universal value of the archaeological and historic environment is expressed most explicitly in the 1972 World Heritage Convention. The convention assumes that the loss of any cultural or natural heritage is an impoverishment to all nations and sets out measures to conserve, protect and present the world's heritage for future generations, through international co-operation and assistance. Within the words of the convention cultural heritage is seen to have outstanding 'universal value' and is seen to be owned by all, for all. Heritage is to be protected in a particular manner under the guidance of international organisations, with formal roles established for the World Heritage Committee, World Heritage Fund, ICCROM, and ICOMOS.

The altruistic notion of conservation as a tool in fostering universal harmony has been used to justify the types of international engagement in this field that has occurred in the later half of the 20th century. I would argue this has often meant that in emphasising the broad universalities and pushing for global conservation approaches, local distinctiveness and difference was ignored or glossed over by the assumed 'universality' of approaches to the past and its conservation. As will be discussed in the second part of this chapter, it is the identification of these problems that has led to the change in philosophical direction of more recent conservation doctrine.

Current international/western conservation theory

In discussing the development of the first ancient monuments legislation in the United Kingdom Sir John Lubbock commented on the selection of prehistoric monuments:

"Surprise has frequently been expressed that we have confined ourselves in the Ancient Monuments Bill to monuments of this character; and have omitted ancient Castles, Abbeys and other similar remains. On consideration, however, it will, I think be felt that medieval monuments require to be dealt with in a different manner. In the first place, the expense would be much greater and ought to be borne partly by local funds and individual liberality. Secondly, as repairs would from time to time be required questions of style and taste would arise, with which no central Commission could, I think, satisfactorily deal; and as to which local opinion ought to be consulted." (preface to Kains-Jackson, 1880; cited in Saunders 1983, 12).

This statement is significant as it illustrates how even in the early development of approaches to the conservation of the archaeological and historic environment, the importance of stakeholders (in this instance landowners) was recognised. By acknowledging both the needs and problems of consulting locally, Lubbock was ahead of the game. From the 1990s there has been a significant shift in conservation philosophy away from the importance placed on the archaeological or historic fabric to understanding, retaining and ensuring participation in a wider variety of 'heritage'.

The questioning of how and in what ways conservation is carried out have been the most significant developments in conservation theory over the last twenty years. Debate has ensued on the unsuitability of the conservation ethic contained within the international/western official charters and recommendations for non-western, non-European contexts. For example critics have argued that the 1972 UNESCO World Heritage Convention was emblematic of the globalisation of western values (Choay 2001), whilst the World Heritage List is viewed as an unrepresentative selection of 'world heritage' and shows a preference for European monuments and/or monuments associated with colonialism (as reflected in the historic fabric) rather than indigenous cultures (Cleere 2000).

These changes in international/western conservation theory are the result of a general move towards acknowledging local difference, otherness and the contextual basis of approaches to conservation (Price 2000), alongside criticism of the top-down Euro-centric approach that has characterised international/western conservation theory (Byrne 1991). The shift from valuing the archaeological or historic fabric to a greater understanding of the scope of 'heritage' value has necessitated a questioning of the suitability of the international/western conservation approach to very different sorts of heritage.

As a result the particular characteristics of the current conservation approaches are: concern with cultural significance as a means to understand a broader range of 'heritage' values, authenticity, participation, sustainability and sustainable development as a tool for poverty reduction, and the notion of intangible heritage. In theory, at least, the conservation debate has moved considerably on from the philosophy of 'anti-restoration' so characteristic of conservation approaches up to the 1990s.

Cultural Significance and Value

Current conservation theory is concerned with the retention of values, alongside the effective and sustainable management of changing values, rather than just the retention of the archaeological or historic fabric. The understanding of cultural significance and values has resulted in a much broader definition of heritage, and redefinition of what conservation means. This widening scope of conservation, through the diversification of 'heritage' and the emphasis on value and cultural significance is seen as a method to encourage wider participation and inclusion within the planning for, and act of conservation.

This shift is perhaps best shown by the incorporation of a greater diversity of values attributed to cultural heritage. Fielden and Jokilehto (1993) state values are the qualities attributed to things, by society, through time. As such values can be complementary, diverse and conflicting. This shift from the tangible archaeological or historic fabric, to the notion of a sense of place and the values associated with a place (its cultural significance) has given rise to conservation theories concerned with the retention of the values and cultural significance of a place.

Emblematic of these developments is the 1999 Australian ICOMOS Burra Charter. This states, "the aim of conservation is to retain the cultural significance of a place" (Article 2.2). The Burra Charter emerged in response to the knowledge that the international/western conservation process had largely ignored the needs of excluded, indigenous communities. The charter was a means to cope with the legitimate claims for participating in conservation and management by defining conservation as a process by which the cultural significance of a place is retained. The charter defined cultural significance as the aesthetic, historic, scientific and social values embodied in the place, fabric, setting and related objects.

The Burra Charter offers guidance for the conservation and management of places of cultural significance through the creation of a three-stage management process; understanding the cultural significance of a heritage place, the development of policy and the management of places in accordance with this policy. In addition the charter defines the processes of conservation, maintenance, preservation, restoration, reconstruction, adaptation and use.

The Burra Charter planning process has been the core tool for conservation planning in Australia for the last 20 years, proving to be highly adaptable to different types of heritage locations (Kerr 1996; Truscott and Young 2000). This value-based management planning process has been adapted to other international/western conservation contexts, and most conservation planning and management theories rely on and draw from the lessons learnt by the Australian experience (Sullivan 1993; 1997, Hall and McArthur 1998), and these approaches have been adopted more widely by international and national organisations concerned with conservation (Avrami *et al* 2000; Clark 1999, 2002; Demas 2002; Mason and Avrami 2002; Palumbo and Teutonico 2002.).

The shifting focus of conservation and the broader understanding of value and cultural significance has generated debate on the public value of heritage and potential tensions between public and expert. This debate places much greater emphasis on the social aspect of heritage, summed up most aptly in recent discussions on the public value of heritage as: "our duty is not just to the places themselves, but to the people for whom they hold value, both today and in the future." (Clark 2006, 99). As a result, the current international/western conservation contrasts with the model of the UNESCO 1964 Venice Charter by placing emphasis not only on the original, historic fabric but also on the culturally rich meanings deriving from a heritage place (Price 2000, Sullivan 1993) alongside the social value of heritage (Clark 2006).

Participation
The last decade of the 20th century saw a growing identification of local needs to participate in conservation as a means to avoid cultural bias (Creamer 1990). As we have seen through the developments of the Burra Charter, heritage is now defined by the cultural significance and value 'people' (rather than specialists) have for a place. As such the conservation planning process is concerned with stakeholder participation, both in the identification of heritage and in the development of approaches to its conservation. As such the majority of the conservation planning documents and guidance are concerned with engaging and registering stakeholder participation within the planning process. These developments follow similar patterns to other planning disciplines concerned with participation and democratisation of decision-making procedures (Glasson *et al* 1994).

The importance of participation has emerged with the acceptance of other histories and indigenous rights to self-determinism (Pwitti 1996), and is signalled by legislative changes such as the Native American Grave Protection and Repatriation Act (NAGPRA), the Vermillion accord and proposed UNESCO World Heritage Indigenous Peoples Council of Experts (WHIPCOE), whilst Article 2 of the ICOMOS Charter for the Protection and Management of the Archaeological Heritage (1990) states the needs for an integrated protection policy whereby active participation forms part of the protection policies. This importance is re-iterated through international charters (such as the Burra Charter) and conservation planning documents used by international and national heritage organisations.

Authenticity
Significant debate has been concerned with the applicability of a western concept of authenticity to contexts worldwide. This is a result of an increased understanding that authenticity is dependent on cultural and social context (Cleere 1995). For example, in reference to the Japanese tradition of rebuilding the Ise Shinto temple Lowenthal concludes, "the concept of conservation thus goes far beyond the acts of material preservation on which Western societies concentrate their efforts" (1985, 385). In this context the 'replaced' building materials would present a challenge to the international/western concept of 'authenticity' which is concerned with the original historic fabric (see section above), whilst the retention of the craftsmen and craftskills (alongside their social significance) indicates a very different understanding of 'authenticity'.

In response to this debate, the Nara Document on Authenticity (1994) was produced. This document rests on the understanding of a broad range of heritage values, and sees the importance of authenticity as a means to, "clarify and illuminate the collective memory of humanity." (Article 4).

"Depending on the nature of the cultural heritage, its cultural context, and its evolution through time, authenticity judgements may be linked to the worth of a great variety of sources of information. Aspects of the sources may include form and design, materials and substance, use and function, traditions and techniques, location and setting, and spirit and feeling, and other internal and external factors. The use of these sources permits elaboration of the specific artistic, historic, social, and scientific dimensions of the cultural heritage being examined." (Article 13 Nara Document on Authenticity).

With the emphasis on authenticity seen through the form and design, materials and substance, use and function, traditions and techniques, location and setting, and spirit and feeling the Ise Shinto temple would therefore be seen as authentic as a result of the retention of craftsmen and skills. This is a significant rethinking of the concept of authenticity and the Nara approach considerably widens the concept of authenticity, and challenges the then conventional approach to authenticity seen within the Venice Charter (1964), and World Heritage Site 'test of authenticity' which placed particular focus on the 'original' fabric (see above).

These changing notions of authenticity have had a significant impact on approaches to heritage and its conservation and management, for example the San Antonio Declaration (1996) links authenticity to identity, history, materials, social value and testimonial value.

Sustainability
Current management and conservation planning approaches are concerned with sustainability. The development of environmental politics characterised by Schumacher's 'small is beautiful ideology' was criticised as advocating 'anti-development' policies and ideologies, and seen as limiting the development potential of Third World countries. In response the concept of 'sustainability' and 'sustainable development', emerged as an ideology concerned with reconciling and balancing equality and environmental issues with development (Dresner 2002). Throughout the 1970s and 1980s the emphasis on 'sustainability' and 'sustainable development' was of minor importance, primarily focussed on natural resources. However from the first wave of green politics that commenced with the 1987 Brundtland report, 1992 Rio Earth Conference, and resulting Local Agenda 21, the notion of sustainability and sustainable development has gained currency in all spheres (Dresner 2002), including cultural heritage. This emphasised the importance of long-term sustainable development and the involvement of local communities (Clifford and King 1996; Solli 2000).

This linkage to sustainable development and emphasis on locality and participation is increasingly seen within an economic context, concerned with both sustainable development and poverty reduction. For example, over the last twenty years there has been a growing awareness of the benefits of heritage and conservation in terms of the wider economic returns and impact on society (such as alleviating poverty through employment and tourism) (Clark 2006). This notion of using conservation as an economic tool for sustainable development alters the values ascribed to the item being conserved, at its very simplest valuing the heritage for its possible economic returns (*op cit*).

Within current international/western conservation theory conservation/management plans are seen as tools for the sustainable management of change through the creation of policies that are feasible as well as compatible with the retention, reinforcement and revelation of values or significance. It is envisaged that a conservation plan will identify what is significant about a place and develop an understanding of how that significance is vulnerable in order to provide for and manage a sustainable future. Within this context sustainability is a contemporary value that recognises current use may pose tensions for the future of the resource.

Intangible heritage

There has been a growing awareness of the needs to protect and retain the world's intangible heritage. The Burra Charter (1999) advocates the continuation and revival of the significant meanings attached to a place (Article 24.2), whilst the Nara Document on Authenticity (1994) saw authenticity defined by use, functions, traditions and techniques. This awareness is stated within the 2003 UNESCO Convention for the Safeguarding of the Intangible cultural heritage. This convention states:

"This intangible cultural heritage, transmitted from generation to generation, is constantly recreated by communities and groups in response to their environment, their interaction with nature and their history, and provides them with a sense of identity and continuity, thus promoting respect for cultural diversity and human creativity."

The Convention categorises intangible heritage as: oral traditions and expressions; performing arts; social practices, rituals and festive events; knowledge and practice concerning nature and the universe; and traditional craftsmanship. This focus on intangible heritage represents a significant shift from conservation of the historic or archaeological fabric to include a much broader understanding of heritage and heritage 'objects' from diverse communities around the world.

As we have seen the conservation debate is dynamic and has shifted considerably over the last twenty years. Section 2.3 will discuss how current conservation theory reflects the concerns of contemporary society.

Current conservation theory and contemporary society

The changing emphasis of contemporary conservation theory must be placed into the context of wider concerns of heritage and contemporary society. This debate builds upon the discussion concerned with the different perceptions of value and cultural heritage outlined above. Further developing the debate concerning the nature of 'difference', 'otherness' and the perception of 'heritage', alongside the conflicts posed by the gap between conservation theory and conservation practice.

By the end of the 20th century broader discussion on the temporal, spatial and contextual basis of conservation had occurred (Denslagen 1993; Theophile 2003). This identified and studied how contemporary society responds to the archaeological and historic environment, recognising that ideas implicit within current conservation theory are not necessarily universal.

Comparative studies concerned with the archaeological and historic environment have observed and highlighted difference in approaches to the conservation of the historic and archaeological environment at a global scale. These differences are generally used to critique different approaches to conservation (typical responses are recorded by Lowenthal 1985 and Stille 2002). These differences have been based on the comparison at a global scale between contexts of use, maintenance and repair, and, contexts of abandonment, conservation and restoration.

The oppositions and contrasts between different approaches to conservation have been extended to envisage a dichotomous relationship between the international/western approaches, and approaches found elsewhere, for example Cleere identifies a restricted concern for archaeological and historical artefacts in, "less-developed societies" (1989, 6).

These contrasting and conflicting approaches to conservation practice are seen as valorised oppositions: use and abandonment, maintenance and conservation, traditional and modern. What is significant is that despite the broader understanding of conservation, and the inclusion of a greater variety of heritage values within the contemporary international/western conservation theory, the observation and valorisation of difference in conservation practice has continued.

Perhaps one reason for this is that despite the development of conservation theory to include a broader class of heritage values concerned with the retention of cultural significance these developments have not necessarily impacted conservation practice. The actual practice and method of conservation has, in many respects, remained characteristic of conservation approaches concerned with the retention of the archaeological and historic fabric.

One reason for the distinction between theoretical and practical concerns of conservation is perhaps that much of the contemporary debate on 'inclusive' approaches has been concerned with conservation planning rather than conservation practice. For example, the 1999 Burra Charter, whilst revolutionary in introducing the idea of value and cultural significance within the planning for conservation, the actual act of conservation emphasises caution and the retention of archaeological and historic fabric. The charter retains the notion, common to all conservation documents, of preserving the life and truth of places and the retention of the aspects that would assist their understanding (Article 3.1 and 3.2). Article 20.2 states: "*Reconstruction* should be identifiable on close inspection or through additional *interpretation,*" whilst "new work should be readily identifiable as such" (Article 22.2), and regardless of the approach the treatment should be reversible (Article 15.2). In this respect the actual conservation practice advocated by the Burra Charter does not significantly alter from that advocated by the 1964 Venice Charter.

As such, despite the emergence of inclusive conservation theories in the later half of the 20th century, the interventionist aspect of international involvement is still evident. The 'correction' of difference remains implicit in the international/western approach to conservation, and Menon (2003) highlights:

"...when these societies set about to conserve their monuments, they adopt Eurocentric norms. The needs for international financial assistance to undertake conservation works and the aura surrounding the UNESCO stamp of approval ensures the adoption of such norms even when viable, traditional alternative exist. What "foreign experts" represent are the "modern" and "progressive" principles of conservation. The desire to align with them is a potent force in developing countries." (Menon 2003, 107).

Underlying this is the danger of only ever paying lip service to widening the remit of international/western conservation through theory, rather than practice. An example of this is found within the Principles for the Conservation of Heritage Sites in China (Agnew and Demas 2004). Within this document the proposed methods of conservation practice and the types of intervention recommended have not shifted away from the focus on the archaeological or historic fabric that typified the earlier conservation approaches specified in the Venice Charter. For example Article 2 of the China Principles document states: "All conservation measures must observe the principle of not altering the historic condition." Emphasising this focus on the archaeological and historic fabric is the statement:

"Physical remains should be conserved in their historic condition without loss of evidence. Respect for the significance of the physical remains must guide any restoration; vestiges and traces of significant events and persons must be preserved." (Principles for the Conservation of Heritage Sites in China Article 21).

The China Principles recommend a cautious approach to conservation *in situ* (Article 18), with preference for maintenance and monitoring as part of a site conservation plan as the most effective means of ensuring preservation. The conservation interventions are stated as: regular maintenance; physical protection and strengthening; minor restoration; and major restoration (Article 28). All emphasise limiting the damage to the original fabric in order to retain the monuments character (Article 30, 31) and shy away from reconstruction, and reconstruction *in situ* (Article 25). When detailing major restoration as a category of intervention the document states:

"It includes returning a structure to a stable condition through the use of essential reinforcing elements and repair or replacement of damaged or missing components...Restoration should, as far as possible, preserve the vestiges and traces of periods judged to have significance." (Principles for the Conservation of Heritage Sites in China Article 32).

This is a problematic statement as it still places most emphasis on the physical appearance (and hence didactic role of both monument and its subsequent conservation), and this is not very far removed from the content and tone of argument used in the earlier conservation dialogue typified by John Ruskin, SPAB and the later Venice Charter. Indeed the accompanying commentary notes for the China Principles focus on the same class of conservation intervention recommended within the Venice Charter, focussing on protective coatings, substances and grouts (Note 11.2), shelters (listed here as 'protective structures') (Note 11.3), alongside minor and major restoration specifying the retention of all components and use of stabilisation (Note 12). The notes go on to specify further additions and replaced components should be marked with the date of replacement (12.3.3). This approach re-enforces a didactic notion of conservation interventions in 'telling' the story of the structure and its subsequent conservation.

Perhaps most problematic with the China Principles is the tone of the document and that despite the progress of international and national conservation dialogues to accept and understand the cultural diversity of conservation approaches (typified by the 1999 Burra Charter, and 1994 Nara Document on Authenticity) this document steps backwards. For example Article 23 states:

"Appropriate aesthetic criteria should be observed. The aesthetic value of a site derives from its historic authenticity. Alterations to the historic condition may not be made for cosmetic purposes or to attain completeness." (Principles for the Conservation of Heritage Sites in China Article 23)

This is a particularly complex statement as the terminology includes two culturally specific terms - aesthetic value and authenticity (for which we have already seen necessitated its own international conservation document). Despite the progress in accepting the cultural diversity of approaches to conservation it remains remarkable that statements, with just a little Ruskinian dogmatic vehemence, such as, "In undertaking repair, it is not permitted to redo decorative painting for new or gaudy effect." (Note 12.1.3) are included: 'gaudy' being as culturally specific a term as 'authenticity'. As such the China Principles document is a striking example of conservation ambiguity.

As can be seen the ambiguity between the concerns expressed in the international/western conservation approaches and what occurs in practice is problematic. Denslagen (1993) states:

"In Europe, with a few exceptions, the Charter of Venice has hardly been taken seriously by restoration architects; it would therefore be somewhat critical were Europe suddenly to insist on the application of principles for European grant-aided restoration of half over-grown ruins of temples in the East. I suspect that in the East what people object to is not so much the sprit of the Charter of Venice as the pedantic tone adopted by Western providers of funds." (Denslagen 1993, 7).

This is not to suggest problems are associated with all of the conservation and heritage charters of more recent years. Indeed some do recognize that the various values and cultural significance of a site may make different approaches to conservation practice appropriate. For example the 1999 Burra Charter does recognise a multitude of different approaches to conservation may be required for a particular place:

"*Conservation* may, according to circumstance, include the processes of: retention or reintroduction of a *use*; retention of *associations* and *meanings*; *maintenance, preservation, restoration, reconstruction, adaptation* and *interpretation*; and will commonly include a combination of more than one of these." (1999 Australian ICOMOS Burra Charter Article 14)

Perhaps more significant is the 2004 INTACH Charter in India. This charter is concerned with finding sustainable approaches to conservation for contemporary Indian society, and does this by addressing and bridging the gap between the international/western conservation theory and indigenous principles and practices of conservation. The charter is concerned with identifying a sustainable interpretation of contemporary heritage theory in order to retain traditional craft skills, preserve cultural diversity and local distinctiveness, and improve social and economic conditions.

The particular focus on conservation approaches for sites that comprise tangible, intangible heritage, their inhabitants and their interconnectivity is particularly significant: "Many unprotected heritage sites are still in use, and the manner in which they continue to be kept in use represents the 'living' heritage of India. (Article 1.1). This document sees the living heritage as one of the key attributes of distinctiveness to be retained and protected from globalisation (Article 1.3), placing particular preference to local, indigenous methods of conservation balanced against international conservation practice and doctrine:

"While the Western ideology of conservation advocates minimal intervention, India's indigenous traditions idealise the opposite. Western ideology underpins official and legal conservation practice in India and is appropriate for conserving protected monuments. However, conserving unprotected architectural heritage offers the opportunity to use indigenous practices. This does not imply a hierarchy of either practice or site, but provides a rationale for encouraging indigenous practices and thus keeping them alive." (2004 INTACH Charter in India Article 2.6).

The significance of this focus on the living heritage allows very different conservation approaches to be accepted as relevant for contemporary society within the Indian context, for example the International/Western approaches to conservation focus on the needs to ensure visibility of repair. For example:

"The legibility of any intervention must be viewed in its own context. If traditional craftspeople are employed then it must be accepted that their pride derives from the fact that the new work is in complete harmony with the old and is not distinguishable from it. Thus, historic ways of building must be valued more than the imperative to put a contemporary stamp on any intervention in a historic building. (2004 INTACH Charter in India Article 3.11.1)

Similarly the focus on the living heritage allows culturally specific approaches to conservation, such as *jeernodharanam,* or regeneration of what decays (Article 3.12). Within this living, Indian context the idea of rebuilding as a conservation approach allows the cultural notion of cyclical perceptions of time. In this context conservation approaches such as restoration, replication, rebuilding and reconstruction are advocated (Article 4.3), alongside conservation using traditional building materials and master craftsmen for certain classes of listed monument (Article 5.13).

The INTACH Charter is significant as it acknowledges contemporary western/international conservation theory, builds upon the developments of the 1999 Burra Charter and 1994 Nara Document on Authentiticy, and interprets conservation theory within the context of indigenous building and conservation practice and traditions.

Summary

The effect of the last twenty years of concern about the universal applicability of an international/western approach to conservation has been to introduce aspects of relativism to current conservation theory. Unfortunately the observation of the contextual basis of conservation is still sometimes contradicted by international conservation recommendations, national legislation, and the remit of international funding and assistance organisations. There are numerous problems associated with the application of an international/western conservation approach to broader, global contexts.

The current international/western conservation approaches can still be criticised for a failure to fully comprehend that conservation activities are contextually derived. For example, the efficacy of applying current conservation theory that seeks to fossilize and make permanent examples of impermanent or transient materials and construction types can be questioned (Markovic 1993).

The characteristics of current conservation theory form the basis upon which the conservation approaches for earthen architecture are assessed in *Chapter 8-16*, and discussed further in *Chapter 17*.

Fig. 1. General view of preparation of earth mix for mudbrick making, Yazd, Iran

Fig. 2. General view of mudbricks stacked on their end to dry, Yazd, Iran.

Fig. 3. General view of mudbricks being made for conservation work, Bam, Iran.

Fig. 4. Stack of dried mudbricks, Merv, Turkmenistan.

Fig. 5. Drying mudbricks, Merv, Turkmenistan.

Fig. 6. Mudbricks with reed and mud plaster to protect from rain, Merv, Turkmenistan.

Fig. 7. Stack of mudbricks, Merv, Turkmenistan.

Fig. 8. Mudbricks with reed and mud plaster to protect from rain, Merv, Turkmenistan.

Fig. 9 & 10. Historic rammed earth wall – '*paksha* made by women' Nurata, Uzbekistan.

Fig. 11. Modern rammed earth wall at the Eden Centre, Cornwall, UK.

Fig. 12. Cob wall, Avebury, Wiltshire, UK.

Fig. 13. Sir Walter Raleigh's Cob house, East Budleigh, Dorset, UK.

Fig. 14. Cob wall, East Budleigh, Devon, UK.

Fig. 15. Cob wall, Gear Farm, Cornwall UK.

Fig. 16. Tall *chineh* boundary wall and tower, Yazd, Iran.

Fig. 17 & 18. Construction of *paksha* wall for conservation work, Bukhara, Uzbekistan.

Fig. 19. *Chineh* boundary wall, Shahdad, Iran.

Fig. 20. *Paksha qala* wall, Old Nisa, Turkmenistan.

Fig. 21 & 22. Turf buildings in Iceland (Photo Malcolm Binks).

Fig. 23. Render removed from *sintch* wall, Bukhara, Uzbekistan.

Fig. 24. Newly constructed *sintch* wall, with fired brick coated in earth render, Bukhara, Uzbekistan.

Fig. 25. Abandoned wattle and daub village pub, Suffolk, UK.

Fig. 26. Restored earth and *gaunch* plaster, Yazd, Iran.

Fig. 27. Variety of earth walls, including *sintch* with hand formed earth balls as infill, Zerafshan Mountains, Uzbekistan.

Fig. 28. Reconstructed drainage spout, Bam, Iran.

Fig. 29. Preparing the earth mix for maintenance of the exterior surface, Khiva, Uzbekistan.

Fig. 30. Preparing the earth mix for maintenance of the exterior surface, Bukhara, Uzbekistan.

Fig. 31. Modern rammed earth wall at the Chapel of Reconciliation, Berlin.

Fig. 32. Wattle and Daub house, Suffolk, UK.

Fig. 33. Earth and cement based renders, Konya, Turkey.

Fig. 34. Tall chineh qala tower, Shahdad, Iran.

Fig. 35. Chineh boundary wall, Shahdad, Iran.

Fig. 36. Modern tall and thin paksha wall for unfinished industrial buildings, Merv, Turkmenistan.

Fig. 37. Sintch building with fired brick infill between splayed timbers, Bukhara, Uzbekistan.

Fig. 38. Eroded earth and decorative *gaunch* plaster, Yazd, Iran.

Fig. 39. Abandoned sintch building, with earth infill, Khiva. Uzbekistan.

Fig. 40. Chineh wall, Shahdad, Iran

Fig. 41. Earth bread oven – tamdyr, Merv, Turkmenistan.

Fig 42. Earth buildings within the Merv Kolkhoz.

Fig. 43. Older *Paksha* building being dismantled prior to new construction, Merv, Turkmenistan.

Fig. 44. Storage and animal houses Merv, Turkmenistan.

Fig. 45. House under construction on a raised earth platform utilising fired brick and mudbrick, Merv, Turkmenistan.

Fig. 46. & 47. Earth burial mound Merv, Turkmenistan.

Fig. 48. Shahriyar Ark Palace wall in 1990s.

Fig. 49. Shahriyar Ark Palace wall in 2004.

Fig. 50-51. Mixing the earth, water and wheat straw mix for maintenance, Bukhara, Uzbekistan.

Fig. 52 Preparing to repair a roof, Bukhara, Uzbekistan.

Fig. 53. Reconstructed drainage spout, Bam, Iran.

Fig. 54. Fallen cement and earth plaster revealing damage to mudbrick wall core, Konya, Turkey.

Fig. 55. Well maintained vertical drain, Yazd, Iran.

Fig. 56. Repaired vertical drain using replacement materials, Yazd, Iran.

Fig. 57. Temporarily affixed drainage pipe, Taklahtan Baba, Turkmenistan.

Fig. 58. Poorly maintained vertical drain, Yazd, Iran.

Fig. 59. Great Kyz Kala 1890 - Photographer: Zhukovsky.

Fig. 60. Great Kyz Kala 2004

Fig. 61. Kepter Khana 1954 – Photographer YuTAKE.

Fig. 62. Kepter Khana 2004.

Fig. 63. Kepter Khana 1954 – Photographer YuTAKE.

Fig. 64. Kepter Khana 2004.

Fig. 65. honeycombing on the Little Kyz Kala, 1924 Photographer: Cohn-Wiener.

Fig. 66. The Little Kyz Kala in 2004

Fig. 67. Palace in Shahriyar Ark 1924 – photographer Cohn-Wiener.

Fig. 68. Palace in Shahriyar Ark 2004.

Fig. 69. Western Gate of Abdullah Khan Kala, 1903. Photographer: Atveladze

Fig. 70. Western Gate of Abdullah Khan Kala, 2003.

Fig. 71 & 72. Stages of abandonment and deterioration, Yazd Old Town, Iran.

Fig. 73 & 74. Stages of abandonment and deterioration, Yazd Old Town, Iran.

Fig. 75. Localised water run off at the base of reconstructed crenellations, Khiva, Uzbekistan..

Fig. 76. Differential erosion of *chineh* lifts, Shahdad, Iran

Fig. 77. Juxtaposition of abandoned and utilised structures, Khiva, Uzbekistan.

Fig. 78. Stages of erosion of earth roofs, Taklahtan Baba, Turkmenistan**.**

Fig 79. Juxtaposition of abandoned and repaired structures, Khiva. Uzbekistan.

Fig. 80. *Chineh* wall with erosion associated with increased humidity from adjacent vegetation, Shahdad, Iran.

Fig. 81. Stages of erosion, Taklahtan Baba, Turkmenistan.

Fig. 82. Differential erosion between lifts and construction 'seams' on *paksha* wall Bukhara, Uzbekistan.

Fig. 83. Differential erosion of surface, Khiva, Uzbekistan.

Fig 84. Materials for backfilling at Aztec, USA.

Fig. 85. Mudbrick capping (and encapsulation) of excavated wall, Gonur, Turkmenistan.

Fig. 86. Collapse of encapsulation work, Gonur, Turkmenistan.

Fig. 87. Encapsulation within *paksha* wall, Bukhara, Uzbekistan.

Fig. 88 & 89. Sandbags used as temporary backfilling materials, Çatalhöyük, Turkey.

Fig. 90. Range of backfilled rooms at Aztec National Park, USA.

Fig. 91. Backfilled room at Chaco Canyon, USA.

Fig. 92 & 93. Partially backfilled rooms at Chaco Canyon, USA.

Fig. 94. Range of backfilled rooms at Pecos, USA.

Fig. 95. Preparing bulk fill for long-term backfilling at Merv, Turkmenistan.

Fig. 96-97. Long-term backfilling at Merv , Turkmenistan.

Fig. 98. Long-term backfilling at Merv, Turkmenistan.

Fig. 99. Trench after long-term backfilling at Merv, Turkmenistan.

Fig. 100. Trench after long-term backfilling at Merv, Turkmenistan.

Fig. 101. Brush wood capping on top of *chineh* boundary wall, Shahdad, Iran.

Fig. 102 & 103. Experimental mudbrick and fired brick capping of excavated walls, Gonur, Turkmenistan.

Fig. 104 & 105. Mudbrick and fired brick capping of excavated walls, Nisa, Turkmenistan.

Fig. 106. Encapsulated *chineh* wall, Yazd, Iran.

Fig. 107. Encapsulated excavated walls, Bam, Iran.

Fig. 108. Encapsulated excavated walls, Rayen, Iran.

Fig. 109 Encapsulation and *in situ* reconstruction, Nisa, Turkmenistan.

Fig. 110 & 111. Encapsulation with replacement materials, and coated in earthen plaster, Shemsh, Iran.

Fig. 112 & 113. Encapsulated excavated walls, Merv, Turkmenistan.

Fig. 114. Encapsulation with replacement materials, Gonur, Turkmenistan.

Fig. 115. Encapsulated excavated walls, Rayen, Iran.

Fig. 116 & 117. Encapsulation cement render, Khiva, Uzbekistan.

Fig. 118 & 119. Encapsulation within *paksha* wall, Bukhara , Uzbekistan.

Fig. 120 & 121. Encapsulation within *paksha* wall, Bukhara , Uzbekistan (2004).

Fig. 122 Encapsulation and reconstruction using harder replacement materials, Khiva, Uzbekistan.

Fig. 123. Encapsulation and reconstruction using harder replacement materials, Bukhara, Uzbekistan.

Fig 124 & 125. Experimenting with earth plasters and geotestile, Merv, Turkmenistan.

Fig. 126. Encapsulation and reconstruction using harder replacement materials, Shahrisabz, Uzbekistan.

Fig. 127. Consolidated and reconstructed 'building 5' Çatalhöyük, Turkey.

Fig. 128. Consolidated 'building 5' Çatalhöyük, Turkey.

Fig. 129. Amended mudbricks used for encapsulation of wall, Pecos, USA.

Fig. 130. Open and eroding trench, Khiva, Uzbekistan.

Fig. 131. Damage to stonework as a result of use of cement-amended mortar, Aztec, USA.

Fig. 132. Eroded and eroding qala, Shahdad, Iran.

Fig. 133. Eroded and eroding defence, Nurata, Uzbekistan.

Fig. 134. Earth mortar plugging drainage gullies, Merv, Turkmenistan.

Fig. 135. French drain cut through archaeological deposits, Merv, Turkmenistan.

Fig 136. Mudbrick and mudplaster altering eroded drainage pattern, Merv, Turkmenistan.

Fig 137. Drainage slope incorporated into backfilling, Merv, Turkmenistan.

Fig. 138. Surface drain filled with aggregate, Nisa , Turkmenistan.

Fig. 139. Surface water collection, Chaco Canyon, USA.

Fig. 140 & 141. Surface water collection for backfilled rooms, Aztec, USA.

Fig. 142. Mudbrick undercut repair, Merv , Turkmenistan.

Fig. 143. Experimental repairs, Gonur, Turkmenistan.

Fig. 14 4 & 145. Mudbrick undercut repairs, Merv , Turkmenistan.

Fig 146 & 147. Experimental house exterior and interior, Çatalhöyük, Turkey.

Fig. 148 & 149. Encapsulation using harder replacement materials, Bukhara, Uzbekistan.

Fig. 150. Encapsulation using harder replacement materials, Shahrisabz, Uzbekistan.

Fig. 151. Consolidated excavated section and wall 'building 5' Çatalhöyük , Turkey.

Fig. 152. Amended mudbricks for encapsulation of wall, Pecos, USA.

Fig. 153. Prayer rags tied onto wire exposed after collapse of cement render used in restoration, Kyz Bibi Mausoleum, Merv, Turkmenistan.

Fig. 154 & 155. Undercut repair, Merv, Turkmenistan.

Fig 156. Experimental house exterior Çatalhöyük, Turkey.

Fig. 157. Painted plaster being conserved prior to removal, Çatalhöyük, Turkey.

Fig. 158. Sogdian wall paintings removed, and conserved on display in Afrasiab museum, Samarkand, Uzbekistan.

Fig. 159. Temporary shelter to allow excavation in shade, Çatalhöyük, Turkey.

Fig. 160. Semi-permanent shelter over the conserved and displayed building 5, Çatalhöyük, Turkey.

Fig. 161 & 162. Permanent south shelter at Çatalhöyük , Turkey.

Fig. 163. Semi-permanent shelter over the conserved and displayed building 5, Çatalhöyük, Turkey.

Fig. 164. Permanent concrete shell structure over Seljuk Palace, Konya, Turkey.

Fig. 165. Shelter formed below visitor walkway, providing shade for visitors to read signage, Nisa , Turkmenistan.

Fig. 166. Permanent shelter over tiled floor, Shahrisabz, Uzbekistan.

Fig. 167. Structure being restored, using mudbricks, Yazd, Iran.

Fig. 168. Restored structure, Yazd, Iran. .

Fig. 169. Restoration in progress, Bam, Iran.

Fig. 170. Restoration in progress, Bam, Iran.

Chapter 8: Backfilling

Format of chapters 8-16

Chapters 8-16 identify the different approaches that have been used for the conservation and management of earthen architecture, for both buildings and archaeological sites. The data from the global review of approaches (*Appendix 1*) and the individual site dossiers (*Appendix 6*) are synthesised. For the different materials and techniques utilised on a site (or sites), the discussion from the site dossier in question is copied, synthesised and, in most instances expanded.

The format adopted for these chapters is to discuss each of the intervention techniques used for the conservation and management of earthen architecture, making reference to the different materials used, and different approaches on historic and archaeological sites around the world. The different approaches are assessed in light of practical impact, relation to current conservation theory, values of earthen architecture and sustainability (summarised in Table 15, *Chapter 17*). The photographs of the different approaches used for the conservation and management of earthen architecture are illustrated from the materials within the appropriate site dossier within *Appendix 6*.

Definition of approaches

The purpose of these different conservation approaches is to alter the environment of a monument and/or site; through protecting and covering the whole site (backfilling, sheltering) or its walls (capping, encapsulation); through maintaining, repairing and strengthening damaged elements (maintenance, surface protection, undercut repair); altering the site or monument setting (drainage), site or monument form (reconstruction, restoration); altering the material properties to add strength and resistance to erosion (consolidation); removing elements from the environment (removal and relocation); and through to 'doing nothing'.

The approaches discussed are:

Backfilling - the replacement of earth after excavation, used either as a temporary measure (such as used between excavation seasons), or a long-term conservation and management approach for excavated earthen architecture.

Capping - the placement of materials more resistant to erosion at the uppermost horizontal wall top - designed so that it is the harder materials which are eroded and/or the harder materials project over the edge of the wall in order to cast water away from the main body of the wall.

Consolidation - the strengthening of earthen materials to make them more resistant to erosion through the alteration of the molecular structure and/or imparting of physical properties to make more resistant to erosion.

Drainage - measures taken to direct or re-direct water away from earthen architecture, these may be concerned with the protection of a single wall, entire monument or site.

Do nothing - not intervening on the historic fabric or archaeological site

Encapsulation - the covering of both the horizontal and vertical surfaces of a monument, wall or trench in new replacement (often harder) materials.

Maintenance - the activities associated with prolonging and keeping a monument or site through preventing deterioration (such as re-mortaring and replastering).

Reconstruction - the rebuilding of a monument or site (or part thereof).

Removal/ relocation - the taking away of a monument or site (or part thereof).

Restoration - the repair and reinstatement of a monument or site (or part thereof).

Sheltering - the shielding of a monument or site (or part thereof) against weather.

Undercut repair - the repair, and reworking of the zone at the wall base most prone to damage from rising water (undercutting).

In a number of instances these approaches are grouped and treated together in order to avoid repetitive discussion of technique, material and site. These are:

Capping and encapsulation

Drainage and undercut repair

Reconstruction and restoration

One problem with the categorisation of different approaches for the conservation of earthen architecture is that often many of these different techniques are used together in order to prolong the life of earthen architecture (for example, capping and undercut repairs, and the blurry distinction between some encapsulation and reconstruction work), and so whilst these broad categories can be discussed individually, there is often a multitude of techniques and materials used in response to the needs posed by the conservation and management of earthen architecture. Whilst the approaches are discussed individually in this chapter, the many and varied character of interventions is reflected in the site dossiers within *Appendix 6*.

Assessment of approaches

The assessment of the conservation and management approaches borrows from the individual assessment made for the site dossiers, and is concerned with particular techniques, materials and approaches in terms of practical impacts, impact of current conservation theory, values of earthen architecture and sustainability. The assessment comprises:

Practical impacts

An assessment of the conservation technique and materials used in terms of survival, deterioration, visibility, and impact on interpretation.

Current conservation theory

An assessment of the conservation technique, and materials used in light of current conservation theories, such as the importance placed on the archaeological or historic fabric, age value, visibility and reversibility, anti-restoration, authenticity, new materials and a role for science and industry, international importance, alongside those more current concerns of cultural and intangible heritage, such as values, participation and poverty reduction. In most instances reference is made to the 1999 Australia ICOMOS Burra Charter.

Values of earthen architecture
The assessment of the conservation technique and materials used with reference to the negative and positive perception of earthen architecture. The values of earthen architecture were discussed earlier in *Chapter 4*. The negative view of earthen architecture is one that sees the material as lacking modernity, associated with poverty, backward and uncivilised, cheap, weak, more liable to destruction, linked to ill health and disease, a last resort, and one with unsuitable language associations. The positive view of earthen architecture values the material as adaptable, aesthetically rich, ancient, autonomous, healthy, locally distinctive, linked to humanity, modern, resistant to environmental disaster, environmental friendly and responsive, and associated with a rich symbolism.

Sustainability
The sustainability of the conservation technique and the materials used is assessed. Sustainability is identified sustainability as a contemporary value linked to the environment, which recognises the importance of the past, *and* how current use may pose tensions for the future of the resource. The broad definition of sustainability encompasses the economic sustainability of the conservation approaches (which is much easier to assess and quantify), the physical and environmental impact of the conservation approaches, alongside broader, more holistic notions concerned with equality between and within generations.

Backfilling

Backfilling is the replacement of earth after excavation. It can be used either as a temporary measure (for example, between excavation seasons), or as a long-term conservation and management approach for excavated earthen architecture. Backfilling is intended to limit the damage caused by exposure to weathering elements by replacing earth removed in the course of archaeological excavation.

The replacement of earth after archaeological excavation has been carried out since the earliest development of archaeology as a discipline (for example, the excavated earth at Silchester was replaced each year to enable crop growth and harvesting (Macaulay 1953)). Work on archaeological sites dug over several seasons was often characterised by the covering over of the excavated parts of a site at the completion of excavation to ensure survival until excavation recommenced in the following year (Barker 1977). Backfilling has been recommended as a suitable conservation and management tool for archaeological sites at an international scale since the UNESCO Recommendation on International Principles Applicable to Archaeological Excavation (1956) and, for earthen architecture, since the late 1960s (for example Torraca *et al* 1972, Balderrama and Chiari 1995 and see recommendations 3rd International Symposium on Mudbrick (adobe) Preservation). Today the types and nature of backfilling associated with earthen architecture are diverse, reflecting the uses of backfilling for temporary measures, long-term measures, and within development contexts for *in situ* preservation.

The documented research has been concerned with developing methodologies for temporary and long-term backfilling, alongside research into backfilling design, investigating issues such as separation (for reversibility), choice of materials for bulk fill, drainage and monitoring. Methodologies developed for backfilling over the last c.15 years at sites such as the Rose Theatre (UK) (Corfield 1996) and over the Laetoli footprints (Tanzania) (Demas *et al* 1996) have been adopted and adapted for use on sites worldwide, and on sites preserving earthen architecture (see papers collated Conservation and Management of Archaeological Sites Volume 6, no. 3 & 4. Special Issue on Site Reburial). Backfilling is considered as one of the only long-term solutions for the effective conservation of excavated earthen architecture (Agnew 1990; Caperton 1990; Feilden 1994; Torraca 1981; Warren 1999), with high profile research carried out at San Diego Royal Presidio, Chaco Canyon, Aztec Ruins (USA); Çatalhöyük (Turkey), and Merv; and other work documented on earthen archaeological sites has occurred in Iraq, Kazakhstan, Kyrgystan, Mexico, Pakistan, Peru, Turkey, USA, Uzbekistan and Zimbabwe. Within the site dossiers backfilling work is documented at Çatalhöyük (Turkey), sites in the USA Southwest (Aztec, Chaco, Pecos), and Merv (Table 5)

Temporary backfilling
Temporary backfilling may be carried out between excavation seasons. The methodology used takes consideration of drainage, ease of implementation, and ease of removal at the commencement of subsequent excavation. Temporary backfilling may also be carried out pending fund-raising and decision-making when excavation is complete.

Temporary backfilling sites
In a number of trenches at Çatalhöyük (Turkey) between-season backfilling has consisted of placing sandbags in trench bases and sides, sometimes on top of a variety of different separator materials (from tarpaulin through to different types of geotextile) (Fig. 88-89). More complex protection was offered with the use of geotextiles formed into 'pillows' filled with a vermiculite/perlite mix in Building 5 (Matero 2000; Matero and Moss 2004). This methodology was developed following laboratory research into temporary protection of the decorated walls (*op cit*) and proved effective in the short-term. Long-term this system has not been retained, and the granular vermiculite/perlite mix now lies scattered across the site.

At Merv a number of different approaches have been adopted for the material exposed in the course of excavation. Excavated trenches have been temporarily protected between excavation seasons using a variety of materials, this includes the use of spoil material as fill and newspaper as a separator layer (in the mid 1990s (*pers comm.* Ann Fuerbach)), the use of spoil material with no separator layer, and more recently, the separation of fill material using geotextile and/or earth placed in sandbags.

Long-term backfilling/reburial
Long-term backfilling/reburial is the infilling of excavated trenches (and/or structures), and may occur at the completion of an archaeological excavation, and also within the context of wider site conservation and management plans, once excavation has been completed for some time. For example, throughout the 20th century many trenches on archaeological sites were left open at the end of excavation, and many archaeological sites around the world have problems related to these abandoned and eroding open archaeological trenches. On these sites backfilling has been used for a number of different reasons, including limiting damage caused through erosion, to cut down the areas that need maintenance, stop drainage problems, limit damage to unexcavated archaeological deposits, and to make the site safer and more legible to visitors. This sort of backfilling can alter the values, significance and appearance of a site, and so decisions regarding backfilling must take into consideration the values and significance of the trench to be backfilled, its vulnerability, the interpretative impacts alongside the long-term sustainability of backfilling.

Backfilling/reburial may also be carried out to fulfil obligations for preservation *in situ* in development contexts, and to fulfil obligations related to repatriation and indigenous populations.

There are numerous materials and techniques that can be employed for backfilling. The materials commonly used for backfilling range from the simple - nothing but replacing the excavated soil, through to much more complex engineered solutions concerned with appropriate separation materials, bulk fill, drainage and monitoring.

Separation - Separation is desirable to distinguish the new material from old, and to make the backfilling work reversible. In the last *c.*15 years separation has normally been provided by some form of geotextile placed at the interface between the trench/limit of excavation and introduced bulk fill. Geotextiles are fabrics made from polymeric fibres which behave as water permeable barriers to prevent the intermixing of soils of different groupings. Normally used industrially for ground engineering and construction projects, there are an enormous variety of geotextiles, all with different properties which affect suitability.

Bulk fill - The material used for the fill is normally the material taken out from the trench, but sometimes different material is transported to site on account of its better properties and/or quantities needed. Earth is either placed loosely or compactly into the trench, or placed within sandbags - retaining the soil and allowing easier removal (if the solution is used for temporary backfilling). In other instances more complex fill material is transported to site, and this includes the use of vermiculite and perlite, and inert sandy materials in use on sites in the USA Southwest.

The backfilling design will specify if it is to be filled in entirely, or partially (retaining the shape and form of the covered wall lines) (Fig. 94). Both of these 'full' and 'partial' solutions have been experimented with at sites in the USA Southwest, Gordion (Turkey) and at Merv.

Drainage - Provision is made for water run-off in the backfilling design either naturally through drainage slopes or through the use of geodrains.

Monitoring - Provision is made for checking the backfilling work, either through simple visual checks, or more complex environmental monitoring. At Merv simple monitoring tools have been used to assess moisture content, pH, and compactness alongside photographic records and visual checks of the backfilled trenches. In contrast much more expensive, high-tech systems of monitoring have been employed in some of the backfilling work carried out on sites in the USA Southwest, such as Aztec and Chaco National Park.

Backfilling sites
A number of sites within the site dossiers provide examples of the different approaches, materials and techniques used for backfilling.

For example, at Aztec Ruins (USA) (*Appendix 6*) two different phases of long-term backfilling/reburial work have been carried out. The first phase was on the southeast range of buildings, in which the backfilling design and materials used were simple, using earth as a bulk fill to partially fill in the excavated rooms, and leave some of wall tops exposed (*pers comm.* Brian Culpepper). The second, more substantial phase of backfilling has been concerned with the west ruin and tri-walled *kiva*. Here the design of the backfilling intervention utilised geotextiles and geodrains, with a bulk fill brought in from off site (Fig. 90-93). To mitigate problems imposed by the bulk fill changing the loads within the rooms, a system of geo-grids was used (Rivera *et al* 2004). An aspect of this reburial has been substantial drainage work, and the installation of these drains has necessitated puncturing through extant walls, alongside damage to the unexcavated sub-surface archaeological deposits. This work is intended to restore the west ruin to its 'unexcavated form', particularly through encouraging the restoration and re-vegetation of grassland (*pers comm.* Brian Culpepper). The decision to partially rebury the exposed rooms was strongly influenced by the perceived long-term financial benefits of reducing the requirements for regular maintenance, whilst allowing public access, and fulfilling the needs imposed by consultation with Native American groups (Rivera, Culpepper, Barrow and Fisher 2004). This work has also required substantial consultation and negotiation with local groups (Nichols 2000).

Similarly at Chaco Canyon (USA) (*Appendix 6*) two different phases of backfilling/reburial work has been undertaken The first phase was the *ad hoc* reburial of rooms exposed in the course of excavation, utilising excavated spoil material as the bulk fill. A second phase commenced from the 1980s, when the National Park Service (NPS) initiated a major reburial programme designed to protect the exposed structures from further deterioration and loss. The motivation of the backfill/reburial project is to cut down on the wall space that requires maintenance and conservation, with backfilling work seen as a means to save money, preserve authenticity (by reducing the need for repair interventions), and fulfil the needs imposed by consultation with Native American groups. The NPS has a remit that requires visitor access and presentation is granted on site, and as a response backfilling/reburial is carried out only on those sites that are not visited, and those that are visited are only partially reburied. From 1991-2003, 16 structures in the park were subject to partial backfilling and drainage works; this amounted to an eighth of the exposed fabric that was most threatened (Ford *et al* 2004). The design of the backfilling intervention utilised geotextiles and geodrains, and due to restrictions on quarrying within the NPS and tribal land, with a bulk-fill brought in from of site (Fig. 84, 92-93). Research has been carried out to establish the best methods for the reburial of the preserved timbers (Ford *et al* 2004; Blanchette *et al* 2004) and monitoring of the backfilling work (Maekawa 2004).

At Merv (*Appendix 6*) where backfilling occurred prior to 2002 it had normally used just with the excavated spoil material and/or sand as fill, and generally with no separator between the excavated and fill materials (although materials such as newspaper were used on an *ad hoc* basis for temporary between season backfilling). From 2002 a number of different methods were experimented with (Cooke 2003; Cooke forthcoming). The materials used included the sieved spoil heap material, which was wetted and compacted as a bulk fill material. The fill was placed on top of, or in front of a geotextile (Terram 500) for separation and reversibility at the horizontal and vertical limits of excavation (*op cit*; Fig 95-100). This method has been used to permanently backfill a number of archaeological and robber trenches either completely (to the original ground level) or partially (leaving the wall line visible, with the provision of drainage slopes). These partial backfilling methods posed particular problems in relation to moderating and accommodating drainage, with the solution experimented with in 2002 (using a mud plaster 'cap' at the limit between backfill material and excavated vertical limits) replaced in 2003 with a more substantial infilling with compacted earth over the backfill material and vertical limit of excavation.

Backfilling assessment

Practical impacts

The practical impacts of backfilling are positive and the approach works well in limiting erosion and retaining excavated earthen materials in its current condition. The approach is less positive as it can sometimes damage the materials it is designed to protect, and can alter the shape and form of the site. Similarly, by making the materials 'invisible' backfilling impacts upon the interpretation and understanding of a site or trench.

At Merv those trenches that have been backfilled seem to last well, and the approach is an effective solution to the conservation and management problems posed by the eroded and eroding trenches. However, at Merv backfilling requires monitoring and maintenance in order to assure its long-term effectiveness. This means that backfilling/reburial does not automatically and permanently reduce the financial burden on an archaeological site, and may actually cost more in the long run especially when complex imported materials are used for separation, drainage and bulk fill.

Alongside practical limits, the ethical suitability of some of the materials used for backfilling/reburial may also require complex negotiation. For example, in the Southwest USA consultation with Native American communities has generated the need to repatriate and rebury material, sometimes as part of the management plans for sites. In these cases the reburial is undertaken without the addition of 'new' backfilling materials, such as geotextiles and/or geodrains (discussions at Santa Fe reburial symposium 2003).

The current backfilling work at Aztec (USA) and Chaco (USA) has created a number of problems associated with drainage in the partially reburied rooms, alongside issues associated with changing the loading characteristics of the rooms (particularly those that were never 'infilled'), and has created an 'unnatural' appearance. The backfilling work has had mixed long-term effects, and when used for those rooms in which timbers had survived (due to the dry climate), the monitoring of the buried timbers showed that the reburial work had caused changes in the environment and caused rapid deterioration (for example see Blanchette *et al* 2004). The decision to rebury rooms has also resulted in changes to the interpretation of the site, for example there are fewer rooms to 'see', and this removes the sense of space and proportion, which had previously aided interpretation and understanding. The surface mounted geo-drains have also had a significant visual impact on the reburied rooms. At Aztec (USA) the final vision of the west ruin as 'restored' to its pre-excavated form through the reburial work is highly problematic, particularly when contrasted with the fully reconstructed *kiva*, which over time as the west ruin is reburied will grow to dominate the site (*Appendix 6*; see below reconstruction).

Long-term backfilling is a controversial solution for the conservation and management of earthen architecture, as it means that the materials are no longer visible. For example, so far no trenches at Çatalhöyük (Turkey) have been subject to long-term, but reversible backfilling, and the absence of permanent backfilling reflects aspirations by the Turkish authorities who prefer sites to be left open under shelter (*pers comm.* Ian Hodder). In contrast the backfilling work carried out at Great Zimbabwe (Zimbabwe) facilitated the reconstruction of the excavated structure *in situ* on top of the backfilled site (Matsikure 2000). Backfilling dramatically alters the shape and form of an archaeological site or trench, in some instances this may benefit site interpretation and legibility (for example the removal of the spoil heaps for bulk fill at Merv), or disadvantage of site interpretation and legibility (as the materials are made 'invisible').

Current conservation theory

Backfilling is recommended within conservation charters, and is a major focus of current conservation research. The need for backfilling is increasingly required within research excavation permit agreements. Often backfilling is recommended when no other conservation solution can be found, for example, the 1931 Athens Charter stated:
"When the preservation of ruins brought to light in the course of excavations is found to be impossible, the Conference recommends that they be buried, accurate records being of course taken before the filling-in operations are undertaken." (Article VI).

Whilst the ICOMOS Charter for the Protection and Management of the Archaeological Heritage 1990, reiterates the role for backfilling by stating:

"the archaeological heritage should not be exposed by excavation or left exposed after excavation if provision for its proper maintenance and management after excavation cannot be guaranteed. (Article 6)

Within this context backfilling has often been viewed as the 'only' option for the conservation of exposed archaeological earthen architecture, with general consensus reached by the 1980 3rd Adobe conference that as a "total protection system" backfilling offered a suitable solution for earthen architecture (Torraca 1980).

Backfilling is viewed as fulfilling the conservation requirements of retention of the archaeological or historic fabric, whilst retaining its 'authenticity' through the reduction of interventions carried out on the archaeological or historic fabric (such as maintenance). Similarly backfilling is reversible, and current research stresses a focus for new materials and a role for science and industry, with the reversibility of backfilling investigated through the utilisation of materials such as geotextile at the interface between the exposed layer and the bulk fill materials. For example, the backfilling work carried out on sites, such as Aztec (USA) and Chaco (USA), has been justified by the needs to fulfil conservation theory by minimum intervention on the historic fabric, and therefore to retain the authenticity of the extant structures.

However, backfilling impacts notions of visibility and alters the age value associations of the archaeological and historic fabric, alongside the cultural and intangible values associated with a site or trench. It is also of note that the reburial strategy at Aztec (USA) will see the west ruin 'restored' to its pre-excavated appearance. By utilising some of the backfilling/reburial work to revert a site to its 'pre-excavated' form we have a hazy distinction between this approach and those approaches seen as more invasive such as restoration and *in situ* reconstruction.

More difficult to assess and understand is the impact of backfilling in changing the values associated with an archaeological site or trench. Current conservation theory would advocate the retention of all of the site's values and significance (Burra Charter Article 1). The decision to rebury sites effectively makes the archaeology and the legacy of the archaeological discovery of these sites invisible. In the former Soviet Central Asia one of the political motivations for archaeology was the presentation of the interpreted view of the past through leaving archaeological trenches open. The legacy of this work is countless archaeological trenches abandoned and left eroded and eroding on archaeological sites. As part of current concerns of conservation and management many of these trenches have been prioritised for backfilling/reburial work (not just at Merv, but also at sites in Uzbekistan, Kyrgystan and Kazakhstan (see Fodde 2003, Hurd 2003). However in planning for the backfilling/reburial of all of these sites it is important to assess how the values associated with the site as a research 'tool' and the legacy of its archaeological discovery can be retained and interpreted to visitors.

Of greater consequence is the impact of backfilling/reburial work on the surrounding archaeological deposits. Backfilling/reburial can damage the historic fabric through excavation below ground to permit drainage. The destruction of material in the course of conservation is problematic, and this type of work can be far removed from notions of minimum impact to the archaeological and historic fabric.

Other concerns are associated with participation and cultural heritage, such as requirements to negotiate and repatriate Native American cultural heritage. For example, the decision to carry out the ongoing programme of 'preservation reburial' at Aztec (USA) and Chaco (USA) is concerned with the repatriation of objects, these objects are reburied incorporated within the fill of the rooms, but to acknowledge sensitivities regarding the use of modern materials, those rooms which are to be reburied with repatriated items are backfilled without the addition of 'new' materials, such as geotextiles (participant discussion at Symposium on site reburial). Backfilling shows how there are considerable and irresolvable problems and paradoxes in carrying out conservation work and meeting the demands of conservation theory.

Values of earthen architecture

Backfilling challenges the perception of earthen architecture as an ancient, durable and universal building material, whilst the materials and techniques used compromise the aesthetic and distinctive qualities associated with earthen architecture.

The decision to rebury the rooms at Aztec (USA) has been justified on account of it reducing the maintenance requirements of the site. This attempt to limit maintenance needs is a significant challenge to the association of earthen architecture with maintenance and renewal activites.

There is also the potential that as backfilling/reburial work seems to be the 'last option', this generates and retains the impression of earthen architecture as an 'un-conservable' material and that the only option is to cover it over and forget about it. The impression given on site that the materials are 'un-conservable' is a powerful image of the perceived limited life span of these materials; this reinforces the negative perception of earthen architecture as a weak building material. The notion of earthen architecture as a 'weak' building material can significantly alter the interpretation and understanding of the past, encouraging visitors to an archaeological site to understand the past as desperate and primitive.

Backfilling/reburial work alters the visibility of the archaeological and historic record and makes it difficult to see and interpret the excavated material. The fact that in many instances the earthen materials are made 'invisible' (as it is the backfilling work that becomes visible) challenges the perception of earthen architecture as a universal building material because the record of use at a particular place and time is no longer visible. This also makes it difficult to understand the material's ancientness, and local distinctiveness. This is significant as by reducing its visible legacy backfilling/reburial can re-inforce the negative perception of earthen architecture as 'unconservable' and impermanent.

Sustainability

The apparent success or otherwise of the current approaches, materials and techniques used for backfilling work impacts on the future of the archaeological and historic sites, and may result in some materials being retained and made visible, whilst others are covered over and made invisible. This approach therefore impacts upon the present generation, as the materials are made 'invisible', this also alters the understanding and interpretation of these sites for future generations. Perhaps the biggest unknown in assessing the sustainability of backfilling is trying to understand how these now 'invisible' sites will retain their significance in the future.

In the past backfilling/reburial programs have achieved negative results due to a failure to monitor and maintain, and this challenges the sustainability of the work carried out. Similarly, the use of geosynthetic materials in the design of the backfilling/reburial intervention presents a number of problems. Despite extensive research and advice, the properties of the numerous geosynthetic materials available for ground engineering applications are complex and not always well understood. This raises the concern that we may in the future be left managing the legacy of the use of geosynthetic materials in much the same way that we are currently managing the legacy of the indiscriminate use of concrete for capping and undercut repair on archaeological sites (participant discussion Santa Fe 2003).

The economic sustainability of backfilling is complex. In the US Southwest backfilling/reburial work is strongly influenced by the need to reduce the financial burden on the NPS to monitor and maintain walls. In the calculations used to assess the viability of backfilling over regular maintenance at Chaco Canyon (USA), the suggestion is made that backfilling would be 5 times the cost of maintenance. If maintenance needs to be carried out every 2-3 years, savings would be realised in 10 to 20 years (Ford *et al* 2004, 181). However these cost calculations do not allow for the maintenance of the reburial/backfilling work, and in the future justifying financial resources for 'invisible' sites may prove to be particularly problematic.

Backfilling can remedy the problems associated with abandoned excavation trenches. The abandoning of excavated trenches resulted in rapid erosion of the excavated fabric, and this has impacted the nature and type of material retained for use for future researchers and visitors to archaeological and historic sites. This is problematic as the non-linear pattern of deterioration associated with earthen architecture means that the most damage to an open trench occurs in the first few years following abandonment (after which time the trench would generally stabilise and erode at a less rapid pace) (Cooke 2002, 2003). The problems of open and eroding archaeological trenches are symptomatic of the type and nature of archaeological involvement that occurred in the past, and in some contexts continues to occur today. Backfilling can be a suitable and sustainable remedy to these problems and works best when done as soon as possible after excavation.

Chapter 9: Capping and encapsulation

Capping is the placement of materials which are more resistant to erosion at the uppermost horizontal wall top. Here it is the sacrificial materials which are eroded and/or harder materials project over the edge of the wall in order to cast water away from the main body of the wall (Balderrama and Chiari 1995). In contrast, encapsulation is the covering of both the horizontal and vertical surfaces of a monument, wall or trench with new replacement (often harder) materials. The purpose of encapsulation is to ensure that erosion occurs to the replacement materials, rather than the historic or archaeological fabric. In addition some materials, such as earth plasters, add cohesion to the original archaeological or historic fabric.

Capping is one of the most common approaches for the conservation of earthen architecture. Similar techniques and materials are used for the capping of earthen walls in living contexts, historic buildings, and on archaeological sites for excavated walls and section baulks (Fig. 103). The materials used for capping range from thatch, vegetation, plastered vegetation mats and timbers (all 'soft' capping), through to harder materials such as ceramic coping and ridge tiles. The main materials used for capping earth walls are earthen materials (mudbricks, plasters and renders), alongside harder replacement materials, such as courses of fired brick, ceramic tiles, or cement (and soil-cement mixtures) (Torroca *et al* 1972).

Encapsulation differs from capping. The latter is concerned with the horizontal surfaces of the upper-most part of walls, whilst encapsulation is concerned with the covering of both the horizontal and vertical surfaces (and to some extent the insertion of new materials to create an even and flat horizontal surface). A variety of different techniques and materials can be used for encapsulation, including earthen materials, either the same earthen material - earthen plasters encapsulated by earthen plaster; or different - a mudbrick wall encapsulated by a placed earth wall, alongside the use of replacement harder materials, such as cement, fired brick, and breezeblock.

For earthen architecture considerable research has been carried out to investigate possible capping materials and solutions. Capping work documented on earthen archaeological sites has occurred in Iran, Oman, Turkmenistan, and USA (with considerable research undertaken at Fort Selden) and Uzbekistan. Within the site dossiers capping is documented in Iran at Yazd, Bam, Rayen and Shahdad, and Nisa and Gonur (Turkmenistan) (and within the other site dossiers combined with encapsulation). In contrast encapsulation is not particularly well represented though it was recorded during the site visits within the study area. Within the site dossiers an overwhelming variety of different materials and techniques are recorded for encapsulation work in Iran at Shahdad, Shemsh, Rayen and Arg-e Bam, and in Uzbekistan at Bukhara, Khiva, and Shahrisabz. (Table 6 & 7).

Technique

For capping the materials put on the wall top project slightly in order to direct water away from the wall, and cast it away from the wall fabric (preferably at least 1 cm to reduce the possibility of rainwater running down the face of the wall, or penetrating further into the structure (Torroca *et al* 1972)). The capping methodology utilised for the wall top is influenced by the nature of the preserved historic or archaeological fabric, for example if the walls represent a regular height and cross section, the capping materials can simply be built onto the wall top, in contrast if the walls are of an uneven height and cross section, parts of the wall will need to be built up to reach an even height (Balderrama and Chiari 1984, 105). In instances where the wall needs to be built up to create an even horizontal layer for the capping materials this solution is not too dissimilar from encapsulation.

The extent of encapsulation varies, and can be the full height of a standing wall and/or the shape of an excavated archaeological site. As with capping, considerable material may be added to the vertical wall height in order to create an even and flat horizontal surface. The disadvantage of encapsulation is the radical alteration to the shape and form of the structure or site.

A number of different materials can be used for capping and encapsulation. These depend on the type and intention of the work carried out. Various different approaches are utilised either:

(1) covering with an earthen material similar to that used in the original construction,

(2) coating in earthen plasters (an approach most prevalent in the documented examples, rather than the visited sites)

(3) covering with an earthen material different to that used in the original construction (such as mudbrick used for a rammed earth wall) or

(4) covering with material different to that used in the original construction (such as cement or amended earthen plaster).

Harder replacement materials were experimented with for capping in early experimental work. In Iraq soil-cement was preferred as a capping material as it was relatively cheap, performed well and was thought to have a suitable appearance (used either in the manufacture of modified mudbricks, or as a modified mud plaster - 'soil-cement') (Torroca *et al* 1972). The soil-cement was used in layers c. 3.5cm of thickness on the top of the walls, however on the test sites the cement and soil cement capping generally showed a tendency for cracking and were not as effective as fired brick (*op cit*). Replacement materials are also utilised for encapsulation. Where earthen and replacement materials are used for encapsulation often the purpose has been to coat the replacement materials in earthen plaster in order to moderate the visual effect.

Capping and encapsulation sites

At Rayen (Iran) the structures in the interior of the citadel have been capped in mud plaster (*kahgel*). As the wall tops were unevenly eroded new materials (mudbricks) have been added to create a level horizontal profile. This approach was used even in those areas in which the original construction was placed earth (*chineh*), which originally would not have been coated in earth plaster. In those instances where the walls associated with a single structure have been coated in plaster, a small mudbrick and plaster arch has also been reconstructed. The purpose of the arch is to indicate the shape and form of the original structure, and this helps to interpret the plan of the site. The *qala* walls have been encapsulated with mudbricks (manufactured with straw inclusions), earthen mortars and earthen plasters (Fig. 108). This work has also involved extensive reconstruction of crenellations at the wall top, and in a number of places this has involved the reconstruction of parts of the city wall to raise the wall higher in order to create a complete circuit of crenellations (Fig. 115). Where this has been carried out the newly constructed wall has been rebuilt using mudbricks regardless of the fact that this replaces the original *chineh* wall.

At Gonur (Turkmenistan) different materials have been used for the encapsulation and capping of the exposed excavated walls and limits of excavation. One part of the excavated palace complex has been partially reconstructed *in situ* through capping and encapsulation within new earthen materials (*Appendix 6*; Fig. 85, 86, 102, 103, 114). The upper layers of the wall top were raised to create a level horizontal profile upon which a thin layer of earthen plaster has been applied (Fig. 85, 86, 114). The mudbricks used here projected slightly over the lower wall fabric to cast rainwater away from the wall. The new capping and encapsulation work was separated from the historic fabric with a covering of thin plastic sheeting (see below for 2004 experiments). Further experiments on the site in 2004 were undertaken to test for the most effective method of capping the excavated material. A variety of different materials and techniques were tried, including modern fired brick and cement mortars; modern mudbrick and earthen mortars; fired brick and curved concrete capping; concrete renders applied directly to the wall top; concrete and earth renders; and the construction of vertical drains to feed into the lined drainage gullies. Even in the small test area the visual impact of this capping work at this site is extreme.

Similar approaches have been used for the encapsulation (and partial reconstruction) of the excavated complex at Old Nisa (Turkmenistan) (*Appendix 6*; Fig. 104-105). A variety of different materials have been used for the encapsulation and capping of the exposed wall tops/tops of trenches. This has included the use of pre-cast cement blocks laid on top of plastic sheeting, fired bricks, and mudbricks. The mudbricks tend to be used to raise the height and increase the width of the walls, with the fired brick placed on top of these as a capping. The result has been to make a walkway around the top of some of excavated complexes, allowing visitors to 'look down' into the excavated spaces. As at Gonur (also in Turkmenistan), the visual contrast between the extant walls that have been left eroding and those that have been encapsulated is extreme.

At Shahdad (Iran) (*Appendix 6*) the excavated archaeological sites have been conserved through the encapsulation of the excavated walls, sections and trenches in mud plaster (*kahgel*). The purpose is to consolidate and add cohesion to the friable eroded earthen surface, and to ensure that erosion occurs in the sacrificial plaster layer rather than the historic fabric.

In Bukhara (Uzbekistan) (*Appendix 6)* encapsulation work used both earthen materials and replacement, harder materials. For example, one stretch of the eroded and eroding historic fabric of the city walls was, at the time of visiting, being conserved through encapsulation within a massive, newly constructed *paksha* wall (Fig. 87, 118-121). This new wall is constructed on top of a concrete footing and lengths of reeds and/or bamboo are placed between each of the *paksha* lifts to add seismic resistance to the structure. At the completion of each lift the substantial zone between the old wall and the new wall has been filled with loose earth. The earth used for this structure has been quarried away from the base of the erosion slope of the existing wall - the quantity of residual artefact material in this debris implied it had been carried out without archaeological excavation or supervision. Replacement materials have also been used for encapsulation, with the exterior of the ark partially rebuilt (Fig. 123, 148, 149). Through the encapsulation work the crenellations at the wall top have been reconstructed. This work involved encasing the ark mound on the western and southern faces (those visible to tourists) completely in fired bricks and cement, on top of this the crenellations have been reconstructed and in places this has been painted white. The effect of this encapsulation work is to create an enormous sweeping buttress-shaped structure. The encapsulation work stops abruptly at the eastern and northern side of the ark, leaving exposed the materials and archaeological deposits that give the ark its enormous elevation.

In Khiva (Uzbekistan) (*Appendix 6*) the defensive wall has been encapsulated using a variety of different materials, including cement and fired brick/tile (see below; Fig. 116-117, 122). Fired brick, flat clay tiles and cement have been used in capping-encapsulation and reconstruction work to create a crenellated parapet on the defensive wall tops. The result has been to create a large drainage slope/buttress on the exterior, designed to create the look and feel of the defended, enclosed town.

In Shahrisabz (Uzbekistan) (*Appendix 6*) sections of the earthen wall have been encapsulated using replacement materials such as fired brick and cement render (Fig. 126, 150). At the wall tops the crenellations have been covered in shaped sheet metal (see below). This has removed the visible traces of the original earthen materrials. The effect of this work has been to reconstruct the shape and form of the appearance of a single phase of the city walls, whilst permitting access to the wall top.

At the time of my visit to Shemsh (Iran) (*Appendix 6*) the earthen caravanserai was in the process of being conserved through encapsulation and capping using a variety of different materials and techniques. The main body of the outer wall and bodies of the inner walls were being rebuilt using a combination of new-fired bricks, and lightweight breezeblocks (Fig. 110-111). These were then covered in a cement amended earthen plaster to generate the 'look' of an earth structure.

In addition the archaeological trench preserving the bathhouse in the residential area of the Arg-e Bam (Iran) (*Appendix 6*) has used a variety of different materials and techniques to retain the exposed materials *in situ*, including fired brick, mudbrick, earthen mortars and renders. These have all been used to cap and encapsulate the excavated material (Fig. 107). In some instances exposed wall bases have been reconstructed *in situ* using newly manufactured fired bricks. The remaining sections of the city wall have also been encapsulated using fired brick, cement, lime mortar, mudbrick, earthen mortars and render. In the Arg-e Bam this work is all finished using earthen renders and lime/*guanch* detailing in order to retain the look and feel of the defences and of a number of structures within the old town. Prior to the 2003 earthquake much of this work had been carried out on the governor's quarter, which on account of its elevated position made an enormous and dramatic impact on the site. The restoration work was also concerned with the stabilisation and encapsulation of the eroding historic fabric within new mudbrick and *kahgel.* This approach was used even in those areas in which the original construction utilised a placed earth technique. The new mudbricks used for the conservation work are made of a similar size to the historic mudbricks, but without the addition of straw: this is to distinguish the new mudbricks from the historic fabric.

At Merv (*Appendix 6*) a variety of different techniques have been used for encapsulation. This included the use of harder cement based materials for the encapsulation of the extant earthen walls adjacent to the Kyz Bibi Mausoleum (Fig. 112-113), the use of harder cement plasters for covering the interior of the Kyz Bibi Mausoleum (Fig. 153), and the experimental uses of earthen plaster applied on top of a variety of different geotextiles on the medieval city walls (Fig. 124-125). In all instances the approaches have achieved mixed results, and in some instances may have contributed to further erosion and loss.

Capping and encapsulation assessment

Practical impacts

Capping and encapsulation have both positive and negative impacts on earthen architecture. The Fort Selden (USA) test wall project documented the effectiveness of capping, here after fifteen years of exposure the 2 capped walls had lost no height, while the uncapped wall had lost 23% of its height, and the cross sections of the capped walls had remained least eroded (Oliver 2000, 64). Similarly encapsulation can work by protecting the archaeological or historic fabric within the materials used to cover the horizontal and vertical limits.

One of the benefits of capping over encapsulation, is that it is a technique used only for the exposed horizontal wall tops, meaning that the wall profile and archaeological sections retain visibility. This is of enormous benefit in understanding and interpreting the historic or archaeological fabric as the didactic evidence of construction type, design and phasing is retained in the wall or section profile.

However, in some instances capping may actually accelerate erosion of the historic or archaeological fabric. These negative impacts are associated with the thinning and weakening of the wall section protected underneath the capping material (see Oliver 2000, 61). Thinning in a zone below the capping is a result of moisture trapped within the wall, which is unable to evaporate through the harder capping material and run-off can be redirected into the wall (*op cit*). In addition, water coming from a soil-cement or cement-amended capping will carry metallic cations producing salts that are deposited in the wall (Warren 1999, 116). These problems can be seen in Turkmenistan at Old Nisa and Gonur, here the capping used cement blocks placed on top of plastic sheeting, this has created an impermeable layer, below which there has been an increased rate of erosion. Similarly in Uzbekistan at Khiva, Shahrisabz, and ark in Bukhara erosion is associated with the encapsulated wall, rebuilt parapet and crenellations, especially where drainage gullies form at the wall tops where the cement renders stop. Even the restoration/encapsulation work within the Arg-e Bam (Iran), which utilised a variety of different earthen and replacement materials and techniques, has created a moderately impermeable barrier below which there has been an increased rate of erosion.

Problems are also apparent with the type and design of the intervention. At Gonur (Turkmenistan) the capping carried out in 2002 used mudbricks and earthen plasters. The width of the capping was not sufficient to cast water away from the walls, and by being just a little wider than the wall, created an unnatural wall profile. The lack of tying-in between the old and new work also resulted in collapse. This means that the conservation work is not structurally sound, and more liable to fall away and collapse. This makes the conservation work reversible, but also unstable, particularly in the event of an earthquake.

After the earthquake the significant damage that occurred in the Arg-e Bam (Iran) was associated with the failure and collapse of those areas that had been capped and encapsulated. This was a result of: (1) the new work being insufficiently attached to the older work, and/or (2) the different qualities and characteristics of the new work exerting different loading characteristics and being less able to withstand movement (associated with being more impermeable and/or having a limited straw content) (see site dossier *Appendix 6*). Given the various different properties and benefits attributed to the inclusion of straw within mudbricks the decision to manufacture new mudbricks without the addition of straw affects the survival of the conservation work and contributed to the dramatic collapse of the conservation work in the 2003 earthquake.

It is particularly problematic when the encapsulation materials extend to the base of the structure or monument at ground level. This results in disruption to the role of capillary breaks and damp proof courses, with moisture becoming trapped. In Bukhara the imposition of the harder, impermeable barriers at the base of monuments has created drainage problems and undercutting at the base of the monuments as a result of capillary action and excess salinity.

The use of harder replacement materials for capping and encapsulation dramatically alters the shape and form of the archaeological or historic fabric, adding difficulties in the interpretation of a site. For example, in Turkmenistan at Old Nisa and Gonur many of the conservation activities have had an enormous visual impact on site; this includes the use of various different concretes, fired bricks, and mudbricks used for capping of the wall tops. The visual clash between these materials adds to the problems of interpreting and understanding. In these instances it is difficult to understand the excavated complexes as the sections and baulks look the same as the excavated walls, whilst those familiar with the use of capping in living contexts may assume that all of the capped areas are walls rather than excavated baulks or sections.

In other instances the enormous scale of the newly constructed *paksha* wall in Bukhara (Uzbekistan) makes it impossible to see and understand the historic fabric it is covering up. This is because it creates a new, very imposing wall with a substantially different size and form when compared with the 'original'. Similarly, in Uzbekistan at Khiva and Bukhara (ark), the effect of the encapsulation work utilising harder materials on the ark and defensive walls has been to create a very imposing monument, in a form that did not exist in the past. As such, encapsulation poses particular problems as the approach makes it difficult to understand a site.

The replacement of the traditional earthen materials with harder, notionally more long-lasting materials generates the perception that the encapsulated monuments are permanent, and therefore require no maintenance. In many places the lack of maintenance of the interventions (either through limited finances or the perception of 'permanence') damages both the restored elements and original fabric. On many of the sites visited, and in particular in Uzbekistan at Khiva, Shahrisabz and the ark in Bukhara the stretches of encapsulated walls were all in need of maintenance.

Current conservation theory

Capping and encapsulation are not advocated within conservation charters as specific conservation techniques. However, capping does fall into a broad category of maintenance activities that may be appropriate for earthen architecture and this category of intervention is mentioned (Burra Charter Article 16). Similarly, the crossover between encapsulation and *in situ* reconstruction and restoration poses a number of problems in relation to conservation theory (see reconstruction and restoration below). In addition, the Burra Charter makes specific recommendations concerning 'new work':

"New work such as additions to the place may be acceptable where it does not distort or obscure the cultural significance of the place, or detract from its interpretation and appreciation." (Article 22.1)

"New work may be sympathetic if its siting, bulk, form, character, colour, texture and material are similar to the existing fabric, but imitation should be avoided." (Explanatory notes to Article 22.1)

"New work should be readily identifiable as such." (Article 22.2).

In this context it can be seen that these conservation approaches can pose many problems as they can distort and imitate a building or site, and the materials and techniques used, mean that it is not always possible to identify them as such. In other respects these approaches *can* result in the retention of the archaeological and historic fabric with only a minimum of interference, but this is at the cost of reducing the visibility of the 'original' work, and dramatically altering the values associated with a site.

In contrast to encapsulation, capping retains the visibility of the vertical wall and section, allowing the historic and archaeological fabric to be seen. This means that the historic or archaeological fabric can be retained with only a minimum of intervention, as advocated by conservation theory. In other instances the more substantial rebuilding required to make a wall or limit of excavation at a similar horizontal level necessitates more substantial reconstruction and rebuilding. This approach would not be advocated by conservation theory, as it imitates the existing fabric and may not be readily identifiable as such (see Burra Charter, article 22.1).

Conservation theory advocates the retention of the visibility of the different building materials and construction phasing in a structure (Burra Charter Article 17). As such capping and encapsulation can allow visual contrast between the materials used and the historic and archaeological fabric. However, balance must be achieved between distinguishing new work and more dramatic alteration to the visual component of the site. For example, in Turkmenistan at Old Nisa and Gonur, the visual impact of the work is enormous, through the encapsulation of the arcaheological fabric the visibility and phasing of the buildings has been altered. This makes it extremely hard to understand and interpret the eroded buildings. On other sites, such as the Arg-e Rayen (Iran), some areas of the extant walls have been coated in earthen plaster, regardless of the original shape and form.

Whilst at Shemsh (Iran) the intervention utilises modern materials, which are subsequently covered in earthen plaster, this creates substantial difficulties in understanding the building.

Materials used for the interventions can (at least in theory) be taken away and the intervention can be reversed. However the utilisation of harder cement-based materials may make reversibility particularly problematic and can result in substantial damage and loss to the archaeological and historic fabric. Similarly the work carried out in Bukhara to encapsulate the remaining stretches of the city wall within the new *paksha* wall is (in theory) reversible. However, the shear scale of the new construction means that any decision to remove the *paksha* walls for future study of the original wall would involve an enormous amount of work, and be on an enormous scale. In addition the infilling of the void between the historic wall and the new *paksha* wall with earthen material makes the process of reversing the work difficult, as the new and old materials blur the interfaces. Conservation theory would perhaps advocate a more small-scale approach to conservation, in order to make the reversal of the work feasible (as advocated by the Burra Charter Article15.2).

Conservation theory would advocate that new work be distinguished from old (Burra Charter Article 22.2). The need for separation is perhaps exaggerated by the nature of earthen materials, where through erosion they can merge together and mix, becoming inseparable. Therefore the separation of the capping work from the archaeological fabric at Nisa (Turkmenistan) and Gonur (Turkmenistan) makes the work (at least theoretically) reversible (as advocated by the Burra Charter Article15.2). However, the damage that occurred to the Arg-e Bam (Iran) after the earthquake indicates the difficulties of balancing the needs and requirements of conservation theory - minimum intervention and reversibility - with the need to ensure that work on site is sufficiently seismic resistant. The collapse of the conservation work on the Arg-e Bam (Iran) has challenged notions of what are acceptable materials and techniques to fulfil the requirements of conservation theory, whilst ensuring work structurally sound.

Conservation theory would seek to retain the existing historic fabric rather than reconstruct and rebuild the postulated missing elements (Burra Charter Article 20). It is particularly problematic that encapsulation often relies on interpretations of the original shape and form of the structure; as such much encapsulation work is similar to *in situ* reconstruction. The effect of the methods used for the encapsulation of many of the city walls in the study area has been to change the nature and form, and to create the defences in a shape and form in which they may never have originally appeared. This has resulted in a loss of the temporal and spatial variation so that all the encapsulated city walls visited seemed very similar.

Values of earthen architecture

Capping and encapsulation have very different impacts on the values of earthen architecture. Capping is an appropriate solution for earthen architecture in historic and archaeological contexts and reflects one of the solutions most often associated with earthen architecture in living contexts. If this approach is well maintained it can reinforce the positive values of earthen architecture, illustrating durability and longevity. When earthen materials are used this approach does present a significant tension between 'freezing' the historic or archaeological fabric in time, and the values of renewal and maintenance associated with earthen architecture.

The conservation work on the city walls in Bukhara (Uzbekistan) is significant as this uses traditional materials and techniques rather than replacement harder materials (as seen elsewhere in Uzbekistan in Khiva and Shahrisabz). The *paksha* creates a very hard and long-lasting building material, although the walls will need maintenance, the work will have a long lifespan. However, the fact that the earth used for the encapsulation may be quarried on site is of concern - in attempting to prolong the life of the defensive wall the archaeological context has been quarried away (although this does quite unintentionally exploit the re-use values associated with earthen building materials).

The use of earthen materials for capping and encapsulation can retain local distinctiveness. However, in Uzbekistan the team involved in the *paksha* construction for the conservation of the Bukhara city walls originated from Khiva (personal communication with work team); as a result they were using a technique perfected in the Khiva region as opposed to the Bukhara region. In other locations around the world (such as Yemen see Marchand 2000) the use of an 'imported' earth building technique has proved problematic, and threatens the local distinctiveness associated with earthen architecture. In addition, the conservation work undertaken on sites in Iran is all very similar and significantly threatens the local distinctiveness of a site or structure.

On other sites the use of harder, replacement materials for capping and encapsulation has dramatically altered the values associated with a site. This approach results in the removal of the earthen elements and changes the soft contours which typify the aesthetics of earthen architecture. The traditional shapes and forms of earthen architecture contrast with the harder, angular shapes created by the use of replacement materials.

In many instances the use of replacement materials reinforces the perception of earth as an un-conservable building material. The use of harder materials for the replacement of the traditional earthen architecture on the city walls has been intended to make these monuments of the past more 'permanent', in contrast to the eroded and eroding extant earthen architecture. In addition the use of harder materials suggests a desire for conservation and management of the site without recourse to regular maintenance. These static conserved walls contrast with the values associated with earthen architecture as a renewable material that relies on community and maintenance for survival.

The manufacture of new mudbricks without straw for the conservation work in Iran (at Rayen and Arg-e Bam) has resulted in the interventions being more prone to destruction and loss. This may reinforce negative views of the material (particularly if the omission of straw inclusions is copied in living contexts). In other instances the use of traditional materials and techniques in conservation has rejuvenated interest in the retention of earth building skills. However, the explicit connection between earthen building skills and conservation, may mean earthen architecture is perceived as associated with notions of the 'past' rather than the 'future', suggesting earth is a building material unsuitable for the expression of modernity.

Sustainability

The different materials and techniques used for capping and encapsulation pose issues in relation to the sustainability of the conservation activities. In some instances the utilisation of harder materials may solve a majority of the problems associated with the erosion and deterioration of the earth walls and excavation limits. In so doing the sustainability of the conservation intervention is assured by retaining the archaeological or historic fabric for future generations (albeit in a substantially different form).

Both capping and encapsulation tend to be limited rather than holistic interventions. For example, at Gonur (Turkmenistan) the conservation work has been restricted to the 'high status' areas rather then the whole site. This is problematice as conservation problems are associated with the remaining unexcavated complexes through natural erosion, damage caused by animals and plants, quarrying and looting. In contrast is the enormous scale of work undertaken on the Bukhara (Uzbekistan) city wall, where long stretches are encapsulated in vast *paksha* walls, irrevocably changing the understanding of the site. Both of these approaches would perhaps be more sustainable if they existed within a more holistic context concerned with broader site conservation and management issues.

A particular problem for the conservation approaches that have used harder cement renders and mortars on the earthen architecture is the near impossibility of removal without substantial damage and loss to the original fabric. This reduces the sustainability of capping and encapsulation interventions by limiting the options available for conservation and management in the future.

Sustainability is perhaps most threatened by the lack of maintenance carried out on the interventions. The replacement of earth with harder materials has generated the perception that these sites and monuments are 'frozen in time' without recourse for maintenance. As with many of the conservation activities there is a tendency for the finances to be "one-off" and suffer from a lack of investment for monitoring and maintenance in the long-term. Deterioration is particularly problematic in the extreme and changing environment in which sites and monuments are located. As such regardless of the type and nature of materials used they will require monitoring and maintenance over time. Maintenance has the potential to increase the sustainability of a site, as it requires a skilled work force to be employed. Maintenance activities associated with an historic or archaeological site may be both sustainable and aspirational (such as being well funded, fully documented, and employing a skilled local work force) or may lack sustainability (being unfunded, undocumented and not contributing to a local economy). As with many of the approaches it is the maintenance activities associated with the conservation work that poses perhaps the greatest possible benefits and disadvantages to sustainability.

These approaches can impact the connection between sites, monuments and locality. The conservation work undertaken at the Kyz Bibi mausoleum at Merv illustrates the linkage between conservation approaches and contemporary society. In the early 1990s a cement render was applied in the interior on top of a lightweight wire frame during encapsulation/restoration work. No maintenance has been carried out to this work, and as a result the cement render survives in a very poor condition with areas of cracked and detached cement, leaving the wire frame exposed. Where the wire is exposed prayer rags have been tied (Fig. 153). The condition of this monument, and the contemporary use of it within Turkmen traditional society and rituals (incorporating the deteriorated conservation work), indicates the tight connection between past, present and future. This may mean that despite the negative practical impact of conservation approaches, the values and significance of monuments and sites as manifested in local customs are sustained for the future.

Chapter 10: Consolidation

Consolidation is the strengthening of earthen materials to make them more resistant to erosion through the alteration of the molecular structure and/or the imparting of physical properties to make more resistant to erosion (Balderrama and Chiari 1995). Consolidants can be used to amend an earthen mix used in conservation, for crack infilling, or can be used for surface protection. Consolidants typically act at a near molecular level through polymerisation, by fixing or inhibiting the capacity for movement between small particles, altering the behaviour of the materials in water, and imparting greater compressive and tensile strength (Warren 1999, 127).

Since the 1960s considerable research has been carried out to investigate possible consolidation materials and solutions for earthen architecture. In the early research consolidation (resulting in 'transformation') was thought to be the only option for earthen architecture (Carter and Pagliero 1966, 68). Consolidation has been carried out as laboratory research and on sites in Guatemala, Iran, Iraq, Italy, Egypt, Peru, Uzbekistan, Turkmenistan, and the USA - with key sites, such as Fort Selden (USA), tending to be the focus of much of the published research on experimental approaches and testing. Within the site dossiers consolidation is documented at Pecos (USA) and Çatalhöyük (Turkey), and within the study area at Samarkand (Uzbekistan) and Nisa (Turkmenistan) (Table 8).

Technique

The various different materials for consolidation are used for:

(1) The make-up of new modified materials that may be used in repair (mortars and plasters etc).
(2) Infill, when it is vital that the soil does not shrink through injection into voids and cracks (and/or the modification of the basic earth mix with consolidants is often carried out to ensure the material does not shrink too much on application).
(3) Application through spraying or brushing on to the surface, intended to strengthen the exposed surfaces, and retain visibility.
(4) Re-adhering fallen or fragile parts.

In a number of instances (such as building 5 at Çatalhöyük (Turkey)) a combination of different consolidation materials and techniques are utilised for the conservation and presentation of the exposed excavated structures.

Materials

A great variety of different consolidants have been used and tested for the conservation of earthen architecture. Some of these consolidants are 'natural' materials, such as plant extracts, whilst others are synthetics.

The enormous variety of these materials include:
(n.b this also includes generic or lab-based research not included in sites table above).

Natural materials

Natural materials used for consolidation include:
Agave juice - this was tested at the Ford Selden (USA) test wall project, and resulted in no colour change but a fairly poor performance as a consolidant (Oliver 2000).
Tuna cactus mucilage – this was used as a consolidant for earthen and lime plasters in laboratory tests (Beas 1993).
Cactus, banana, locust beam tree - used as consolidants on test walls subject to simulated weathering and erosion (Neumann and Mehta 1987).
Linseed oil - used as a coat and as an amended earth render, proving to be generally successful at Ford Selden (USA) (Oliver 2000), and other Southwest USA sites, where it was mixed 1:2 solution in mineral spirits (Taylor 1987). Linseed oil is often used in living contexts for the treatment of earthen walls and floors.
Gypsum (juss), lime (hydrated and hydraulic) - added to the earthen mix when preparing the earth to make harder, the use results in a colour variation of the earthen material (see Table 1).

Synthetic materials

Synthetic materials utilised for consolidation include the use of:
Asphalt - makes water resistant, and is used to amend earthen plasters (Oliver 2000; Taylor 1987). Asphalt has a long history of use as a consolidant for earthen architecture (Table 1); asphalt tends to alter the colour of the surface.
Bitumen - bitumen applied to earth surfaces, and generally penetrates the outer layers of the earth, adding a degree of water repellence, whilst retaining the flexibility and longevity of the surface. Bitumen alters the colour of the surface, and may trap moisture within the wall fabric.
Cement - used to amend the basic earth mix through the formation of crystals between the clay particles (used for foundations with 1-3% Portland cement; capping with 5 to 10% Portland cement; and for lintels and other load-bearing components for repair (Warren 1999, 116)). Cement amended earths are also used for crack infilling.
Latex – used for surface treatment at Fort Selden (USA), where it achieved poor penetration into the earth, but did produce distinctive repairs in accordance with current conservation theory (Oliver 2000).
Methyl methacrylate/ethyl acrylate resin - most commonly used as an additive to earthen mortars used for repair in the Southwestern USA (Hartzler 1996; Oliver 2000). The consolidant increases the moisture resistance of soils, and imparts greater resistance to freeze-thaw damage. The use is problematic due to the propensity of the material to change colour and trap moisture (*op cit*).
Isocyanate monomer - utilised for surface treatment by the Institute of Archaeology of the Academy of Sciences of Uzbekistan (commencing in 1967), this class of consolidants has the potential to increase strength, without changing the appearance of the consolidated wall (Abdurazakov 1986). The polymerisation reaction can occur under 'normal conditions' (in the sunshine) without recourse for complex and expensive infrared heating of the surfaces.

PFA (pulverised fuel ash) - used as a filling material for gap-filling and block-forming at a molecular level (the PFA particles are globular and move through and fill small porous voids). The materials have a tendency to alter the colour of the consolidated historic or archaeological fabric.
Polyvinyl acetate - used as a surface treatment, especially in the Southwest USA. Problems associated with the relative impermeability of the consolidant resulting in trapped moisture within the earthen fabric (Taylor 1987b; Oliver 2000).
Silicone resin – used as both a modified render and as a surface coat at Fort Selden (USA), the use is problematic as it has a tendency to create a harder impermeable layer below which there is an increased rate of erosion (Oliver 2000).

Silicates - the use of this broad range of consolidants is intended to deposit silicate crystals, by the use of organic salts which decay by polymerization to leave the silicates in place and gives to a regular matrix of silica within the clay particles. The compounds within the soil will break down to leave inorganic components (silicates) deposited within the earthen material (Warren 1993). Silicates were first developed for stone conservation in the mid 20th century (see Wheeler 2005) and have been adopted and tested for earthen architecture since the 1960s. The benefit of this group of consolidants is that they can be used in combination or in advance of other approaches, materials and techniques. The disadvantage of this group of chemicals is the irreversibility, relative cost, and difficulties of application.

Ethyl silicate - used for surface protection to vertical surfaces or on steep slopes. The use of ethyl silicate is perhaps the most numerous of the documented approaches concerned with the consolidation of earthen architecture. Silica esters react with the clay particles forming a 3-dimensional network of silica bridges, which increase the water resistance of the material (Balderrama and Chiari 1984). Ethyl silicate reacts with water in the presence of the catalyst and the polymerization can be either rapid or slow (with advantages recorded for a slow process). The material is applied through boreholes or spraying, which alters the depth of penetration and the dispersal of the consolidant. The surface maintains its porosity and internal moisture can evaporate, with the benefit that further treatments can be applied in the future. On archaeological sites the treatment needs to be performed as soon after excavation as possible (as silica esters do not have gluing properties) (Balderrama and Chiari 1984, 105). However the protection it affords to horizontal surfaces is insufficient to cope with the erosive action of heavy raindrops and rainwater (Torraca *et al* 1972, 281). The treatment is not reversible contrasting with the properties of consolidants advocated through conservation theory.

Methyl silicate - reduces the speed at which weathering takes place, and does not change the colour, or characteristics of the treated surface. The disadvantage is the expense, and length of time required for application.

Potassium silicate - creates an interlocking crystalline network with rigidity in the clay plates. The material has been used on test sites in China, particularly on sites with problematic Montmorillonite clays (where ethyl silicate is inappropriate) (Zuixiong 1990). The use of potassium silicate is particularly problematic as it has a tendency to create an impermeable surface, below which there is an increased rate of erosion (Taylor 1987b; Oliver 2000).

Synthetic resins - this broad class of materials is used to modify new materials used in repair, as infill or for surface treatment.

Acrylics - a great variety of different materials have been experimented with and utilised as a surface treatment, but generally found to be ineffective as they form a film on the surface, the characteristics of which are different from the untreated parts, and has a tendency to exfoliate and detach, causing more damage to the surface (Balderrama and Chiari 1984, 105). Acrylics have also been used for injecting into cracks on the surface (Taylor 1987; Oliver 2000). The use of these materials generally alters the colour and appearance of the consolidated walls and by limiting moisture transfer through the surface results in trapped moisture (Oliver 2000).

Epoxies and polyurethane - used as a surface treatment but generally ineffective as they form a film on the surface, the characteristics of which are different from the untreated parts, this film then has a tendency to exfoliate and detach, causing more damage to the surface (Balderrama and Chiari 1984, 105).

Sites
During the first phase of archaeological activity at Çatalhöyük (Turkey) (1960s) all of the wall paintings were removed to Ankara Museum (*Appendix 6*). A variety of techniques were used: *strappo* (detachment of the paint layer alone), *stacco* (detachment of the painted surface including the underlying plaster surface), or *stacco a massello* (removal of entire walls). In addition polyvinyl acetate was used in the field on the very poorly preserved walls prior to removal. Paintings were detached as blocks and faced up using PVA with and without Japanese tissue and linen (Matero 2000, 79). The detached wall paintings were prepared for display and consolidated using polymethyl-methacrylate.

During the second phase of activity at Çatalhöyük (Turkey) (1992 - today) a variety of different methods have been tested and used for surface treatment and crack infilling. The purpose is to allow for the chemical consolidation of the material exposed in the course of excavation, and its retention under shelters (Fig. 127-128, 151). In the current phase of engagement at Çatalhöyük similar methods for consolidation of the exposed walls *in situ* have been used as those used for the consolidation of the lifted wall plasters in the 1960s. This has again involved the use of PVAs to stick flaking and delaminating wall plasters back together again. The methods of application have been experimented with, and have generally found the best results from spray application. In addition thinner and larger cracks have been infilled using modified grouts and mortars. The excavated walls have been consolidated using a combination of the following techniques and materials - acrylic emulsion to re-adhere delaminated plaster, natural hydraulic lime grouting injected into fill thin cracks and mortar to fill larger cracks. The success of these approaches in the sheltered 'building 5' has encouraged similar use across the rest of site (*Appendix 6*).

At Pecos (USA) (*Appendix 6*) an extensive variety of different materials and techniques have been used for consolidation, encapsulation and stabilisation. The excavated and exposed masonry has been coated in consolidant. The exposed parts of the partially backfilled pueblo complex have been treated with E330 acrylic modified mortar to assure longevity of the masonry work. Other materials used for consolidation include the use of Rhoplex E-826 on the remaining adobe walls, sometimes used with an overcoat of water repellent. Additionally a wall was pointed with an amended earth made with Rhoplex E-863 (Hartzler 1996; Fig. 129, 152).

The extant defences of Afrasiab, Samarkand (Uzbekistan) were used from 1967, as one of the testing grounds for the Institute of Archaeology of the Academy of Sciences of Uzbekistan, for the investigation of the uses of polymers and monomers for the conservation of earthen architecture (Abdurazakov 1986) (*Appendix 6*). In Samarkand these materials were used on 630m² of the defensive walls (used after some rebuilding and supporting of the wall with new mudbricks, *paksha*, and earthen plasters) (Reutova and Shirinov 2004). The Sogdian wall paintings preserved *ex situ* in the Afrasiab museum have also been consolidated using a variety of unspecified chemical consolidants and adhesives (see removal/relocation below).

Consolidation assessment

Practical impacts

The use of consolidation for the conservation and management of earthen architecture poses considerable practical impacts for an historic or archaeological site. These are associated with the success or otherwise of the intervention, alongside the visual impact of the intervention, often altering texture and colour.

Much of the work concerned with the testing of chemical consolidants tends to yield negative results. This is because many of the materials that have been tested with some success in a laboratory have a tendency to prove less effective when used on site. This is associated with *in situ* environmental deterioration and erosion, and the changing properties of archaeological or historic materials (such as the washing out of fines), when compared with the newly manufactured earthen materials which laboratories tend to experiment with. In addition, for those sites in which tests have been carried out monitoring and maintenance has proved particularly problematic in the long-term (Torraca *et al* 1972, Chiari 1990a and 1990b). For example, the various different approaches to chemical consolidation of the earthen walls on Afrasiab, Samarkand (Uzbekistan) have generally proved to have a limited success when used *in situ* (although they normally produce very good results in the laboratory or museum) (Reutova and Shirinov 2004). During *in situ* testing phases problems were noted with the insufficient penetration of the wall under consolidation, problems with the horizontal, wall top application, and costs. Long-term the results have proved problematic on account of the limited longevity of the chemicals used, and the need for work to be monitored and maintained.

To avoid the problems associated with erosion a number of the sites have consolidated wall paintings preserved *ex situ* in museum contexts, or have been consolidated below permanent shelters. In these contexts chemical consolidation can prove to have very successful results (especially building 5 at Çatalhöyük (Turkey)). Experience shows that despite the apparent success of such methods for the conservation and consolidation of earthen architecture *in situ* under shelter, the methods often fail due to a lack of maintenance (of either the conservation work itself or the shelter structure), or the negative impact of further work in an area that extends beyond the limits of the conserved area. A similar situation can be seen at Çatalhöyük where, despite the success of the conservation work in Building 5 problems associated with shelter maintenance and wider site issues, such as drainage, threaten the otherwise successfully consolidated materials.

With all of the materials utilised and tested for consolidation, problems are posed as the materials used may have different properties and co-efficients than the earthen materials they are consolidating, and so may result in accelerated differential erosion, such as the shearing of the consolidated surface (see Oliver 2000). Problems are also associated with the use of chemical consolidants as the chemicals may migrate and breakdown through the structure. For example, the addition of cement as a modifier is problematic as it can be incompatible with earth, and can produce soluble salts. They can also result in significant colour alteration, increased brittleness, and changes to texture. One aspect particularly noted is the very different texture and erosion qualities of cement amended mudbricks and mortars, for example not forming soft eroded corners (Oliver 2000, 63). The use of cement as a cladding or capping material is also problematic as it limits water movements, the resultant excess trapped moisture results in failure due to the saturation of the earthen wall, and can cause cracking (Oliver 2000, see Fig. 131). In addition attempts to remove cement renders are problematic, and may result in further collapse.

In other instances the use of chemical consolidants changes the appearance of a site. For example, at Pecos (USA) those areas reconstructed or encapsulated with acrylic modified mudbricks and mud plaster look unnatural, having a shiny, glazed character. Many of the concerns of utilising consolidants have been associated with resultant colour change. For example, the factors used in the assessment of the effectiveness or otherwise of the consolidants at the Fort Selden (USA) test walls were the propensity of the consolidants to change the colour of the wall. In other instances the materials used for consolidation have a tendency to create a film, or impermeable barrier/layer, which limits moisture transfer from the wall and so creates a zone more prone to erosion below the consolidated surface layer.

Experimental approaches for the consolidation of earthen architecture have generally achieved negative results. One reason for this is that knowledge tends to be concentrated in relatively few specialists, and there are problems with the documentation and transfer of these materials and techniques to other people and other sites. The application of the complicated chemical consolidants is expensive, and a number of the materials utilised have been superseded and/or banned on health and safety grounds. Even on a single site using the same methodologies problems arise between different teams and periods of involvement, for example it is unfortunate that the methods tested at Çatalhöyük (Turkey) were not fully documented, monitored or maintained, and assessing the success or otherwise of the conservation methods now is problematic.

The problems associated with the efficiency and efficacy of utilising consolidants include poor penetration and subsequent shearing of the surface, changes to the colour of the historic fabric, long-term efficiency, costs and health implications. As a result regardless of the use of consolidation, maintenance remains a vital aspect for the conservation of earthen architecture (Chiari 1990).

Current conservation theory

Consolidation is recommended within conservation charters, and it remains a major focus of current conservation research. The 1931 Athens Charter states:
"(they) approved of the judicious use of all the resources at the disposal of modern techniques and more especially of reinforced concrete.

They specified that this work of consolidation should whenever possible be concealed in order that the aspect and character of the restored monument may be preserved.

They recommended their adoption more particularly in cases when their use makes it possible to avoid the danger of dismantling and reinstating the portions to be preserved." (Article IV).

Within this context the different materials and techniques utilised for consolidation were designed and tested in order to fulfil the requirements of current conservation theory, such as the need for visibility and reversibility, the emphasis on new materials and the role for science and industry.

However by the later half of the 20th century some of the practical problems of utilising consolidants gave concern over the appropriateness of the use of consolidants. For example the Burra Charter states:
"In some circumstances modern techniques and materials which offer substantial conservation benefits may be appropriate" (Definition for Article 4.2).

However:
"The use of modern materials and techniques must be supported by firm scientific evidence by a body of experience." (Notes for Article 4.2).

In this context some work associated with earthen architecture has been concerned with the definition of principles for the use of consolidants for the material, such as the Adobe 90 preprints stating the '*ICOMOS Principles for consolidants: (Applicable for synthetic materials aimed at hardening the surface and those aimed at deep penetration)*', setting out and stating a number of requirements, such as:

1. It should be removable, i.e. reversible.
2. It should be available at a cost level which allows its use on a broad scale
3. It should be straightforward in application and not damaging to the environment or the applicator
4. It should be capable of being carried into earths in solution in a medium which will not damage the structure and which will disperse without danger or environmental damage.
5. It should not cause any colour change or form film on the surface
6. It should diffuse into the soil progressively rather than forming a precise boundary.
7. It should resist or at least be unaffected by the capillary movements of water, and should be hydrophobic.
8. It should resist the pressures of crystallisation of salts and the pressures caused by freezing of water.
9. It should be permanent, being neither evanescent nor affected by ultra violet light, oxidation or other forms of decay.
10. It should allow water to move through the material both as liquid and vapour leaving the pores in the material open.
11. It should add to the mechanical strength of the material without inducing brittleness.
12. It should be stable and transportable in whatever from it may be available in prior to application.

In addition parameters for the use of synthetic consolidants and the conservation of earthen architecture, developed by the Institute of Archaeology of the Academy of Sciences of Uzbekistan, are stated as:

1. Consolidating agents must not distort the structures' original colour and texture.
2. They must penetrate easily into the thickness of material, providing durability of conservation.
3. Consolidating agents must be resistant the external climatic factors (Abdurazakov 1986, 83).

These requirements and parameters advocated by conservation theory have proved very difficult to achieve, and a great deal of time and money has been expended in the research and development of appropriate consolidation materials and techniques designed to meet the needs of conservation theory. It is problematic that many of the materials and techniques are developed in laboratory environments, and whilst they prove effective in the laboratory they have very variable effects when used on site, or the consolidants and approaches may fulfil a number of these conservation aspirations but fail with others. For example, at 'Building 5' at Çatalhöyük (Turkey) the chemical consolidants do not cause any colour change or form a film on the surface, however they are not reversible. Similarly the consolidation work requires revisiting and reworking, and without regular maintenance cannot be 'permanent'.

Despite the problems associated with the use of chemicals for the consolidation of earthen architecture they remain extremely popular. The use of chemical consolidants fulfils the notion of minimum intervention on site. The use of consolidants for surface treatments can retain the visibility of construction details and stratigraphy, which is an important aspect of site presentation and interpretation. 'Building 5 'at Çatalhöyük (Turkey) is conserved and presented with a minimum of visual interference, with the consolidants retaining the shape and form of the excavated areas.

In other instances the use of chemical consolidants alters the values associated with an archaeological site, and can limit the research potential of an excavated feature. For example at Çatalhöyük (Turkey) the approach is adopted for the conservation of the first exposed layer of wall plaster. However, the irreversibility has resulted in tensions between the conservation and archaeological teams as the conservation approach is perceived as limiting the investigation and understanding of the complex micro-stratigraphy of successive plaster layers (discussion at Çatalhöyük, see *Appendix 6*). As we have already identified current conservation theory would advocate the retention of all of the sites values and significance (Burra Charter Article 1), not just the 'first' exposed layer.

Values of earthen architecture

The use of consolidants for the conservation of earthen architecture challenges the perception of the material through the alteration of the material properties, appearance, aesthetic qualities and value of the material as 'recyclable'.

Some consolidants change the appearance and aesthetic qualities of earthen architecture, making the historic or archaeological fabric appear 'harder' and 'shinier'. For example, one effect of the materials used at Çatalhöyük (Turkey) is a change in the colour and texture of the structures, making them seem glossier and more compact. In other instances the fact that surface treatments can retain the visibility of the construction detailing and design presents the unrealistic view of the walls always looking this way, whilst in reality they would have been coated in earthen plasters. The use of surface treatments and consolidants presents the incorrect idea that these unrendered earthen walls would have looked like this in the past, altering the aesthetic values of the conserved wall and changing our understanding and interpretation.

However, in some instances these approaches are suitable. An explicit attempt has been made at Çatalhöyük (Turkey) to preserve material in its excavated form, so in Building 5 surfaces, sections and blocks of stratigraphy have been treated with chemical consolidants. The clinical, static approach to conservation preserves both wall lines and trench sections (preserving stratigraphy). This makes clear the distinction between the excavated material as a static (and durable) record, and the living legacy of earthen architecture through maintenance and replastering.

Similarly the various different approaches to conservation at Pecos (USA) and Afrasiab (Uzbekistan) have been intended to retain the historic earthen fabric forever, limiting the recyclability of earth as a building material; this also results in the material no longer being able to 'breath'. This presents us with a paradox that these methods can assure the longevity and perception of durability of earthen architecture, but at the detriment to the positive associations of earthen architecture as a breathable, recyclable material.

Unfortunately the limited success of the chemical consolidants has reinforced the perception of earthen architecture as unconservable. This is because developments in science and new technology have been perceived as failing to provide a solution. As such we assume there is nothing more that can be done to assure the conservation of earthen architecture (see *Chapter 17* for further discussion).

It is significant that structures constructed or conserved with materials perceived to be harder and longer lasting have performed less well, and survive in a poor condition when compared with those constructed or maintained with unamended earth. This is particularly well illustrated by the variable condition of the experimental cottages constructed as Amesbury, Wiltshire (UK) (*Appendix 6*). Here in the 80 years since construction it is the experimental earth structures that survive in a much better condition than the concrete (or amended concrete) structures, similar patterns can be found around the world. The fact that it is the unamended earthen structures that survive better and have much greater durability is rarely highlighted and this re-enforces the perception of earthen architecture as lacking durability.

Sustainability

The use of consolidants can challenge the notions of sustainability of the conservation of the archaeological and historic fabric. In a number of instances, such as Çatalhöyük (Turkey), the use of a combination of chemical consolidants, alongside retention under shelter has proved, at least partially, successful enabling the materials and site to be visited, understood and interpreted both now, and in the future.

In other instances the use of consolidation is not particularly sustainable. Indicative of this is the fact that the acrylics used to modify the earthen materials at Pecos are no longer available for use in conservation work (*pers comm.* Pamela Jerome). Similarly some of the methods for chemical consolidation used at Çatalhöyük (Turkey) must be questioned in terms of long-term sustainability, for example although the consolidation work that has occurred in 'building 5' seems to have worked and lasted well for the initial period, areas are now deteriorating and require maintenance. As a result of the limited documentation of the materials and methods originally used, the replication and maintenance of this work is problematic. This makes planning for maintenance, assessing long-term success, and planning to replicate the same material and techniques across the site very difficult and unsustainable. This re-enforces the need for accurate and appropriate documentation and maintenance to assure the sustainability of the conservation approach.

Similarly many of the more complex chemical materials used pose problems, as they tend to be developed within laboratories, and tend to have only a limited effectiveness when used on site. Many of the problems associated with the survival or otherwise of the consolidated walls are associated with the lack of management of wider issues, such as drainage, protection to wall tops, and protection to wall bases. This emphasises that whilst solutions can be postulated and experimented with, if they are used on archaeological sites and historic buildings without the necessary needs for management, maintenance and documentation they will not be successful and will lack sustainability in the long-term.

Problems are associated with the specialised nature of the application of many of the consolidants. The knowledge base for these experimental approaches is relatively narrow and relies on outside specialists for use on site, disempowering those locally concerned with site conservation and management. Controversy surrounds the use of chemical materials, and the perception has emerged that chemical consolidation is the only 'correct' way to proceed with the conservation of earthen architecture. Highlighting this tension, is the discussion of the conservation of a mudbrick wall uncovered during the excavation of the Crypta Baalbi, Rome (Italy), here more traditional unmodified earth and lime materials were used rather than complex chemical consolidants, the comment is made:

"This work has been presented in order to emphasize how a "light" approach in conservation can produce "heavy" results, without the use of tons of synthetic products. We deliberately chose to move away from the now almost "traditional" type of conservation that calls for massive use of synthetic products, applied to incompatible substrates. This approach signals the mutation of our profession as "conservators" (of form and materials) into that of "transformers" (of original artifacts into "healthy objects")." ... "This line of work, apart from its obvious theoretical value, has important practical implications. The "official world of conservation", composed of the few countries and institutions that benefit from advanced technology and that are viewed as models, are rapidly outdistancing the others. Thus dramatically increasing the gap between this elite and the rest of those responsible for cultural property. The price is paid, not only by the heritage itself but also by the "followers", who suffer in the form of professional frustration, for not being "technologists", and who often, as a result, abandon trying to work with the means at hand and, worse still, abandon the traditional manuals of care and maintenance that are themselves part of the world cultural heritage." (Nardi 1987b, 76- 77).

Nardi re-enforces the notion that sometimes it is the local knowledge and simple materials and techniques rather than the complex consolidants for conservation interventions that prove to have the greater long-term sustainability.

Chapter 11: 'Do nothing'

Not intervening on the historic fabric or archaeological site is both a passive and active response to the problems posed by the conservation and management of earthen architecture.

There are countless sites around the work in which this approach has been adopted. Within the study area this approach was recorded on archaeological sites (in Uzbekistan at Nurata, Samarkand, Khiva, and Shahrisabz and in Turkmenistan at Jeitun and Merv) and buildings (Shahdad (Iran) and Taklahtan Baba (Turkmenistan)).

It is important to note that the idea of 'doing nothing' is different to 'not having any other option' or not bothering to do anything with it. There are distinctive differences between sites in which this approach is adopted, for example:

- the building or archaeological site may be documented or undocumented;
- the building or archaeological site may be within official or unofficial policy and management contexts (land ownership etc);
- 'doing nothing' may occur at any stage between abandonment, deterioration and formation and deformation of archaeological sites (e.g. to buildings, ruins or archaeological sites);
- 'doing nothing' may occur at any stage after disruption between abandonment and deterioration, and formation and deformation on archaeological sites (e.g. after older conservation work or after archaeological excavation);
- 'doing nothing' may be an active (decided upon) or a passive (no practical alternatives, no economic or political incentives) response.

n.b the 'do nothing' approach is recorded on numerous archaeological sites around the world – but does not feature within the documented approaches to the conservation and management of earthen architecture, and as a result is not analysed in the same manner as the other approaches.

At the unexcavated archaeological site at Nurata (Uzbekistan) the ramparts and defensive structures have received no conservation interventions, and the unexcavated archaeological site has been left with no provision made for its conservation and management (Fig. 133). Similarly the unrecorded building complexes comprising the *qalas* at Shahdad (Iran) have by-and-large been left abandoned (Fig. 132). The effect of leaving the historic fabric untouched is that there is a substantial amount of eroded and eroding earthen architecture and in most instances these structures have already lost their protective roofs. Similarly at Taklahtan Baba (Turkmenistan) a majority of the Maddrassah complex and associated structures have fallen into disuse and have been abandoned (Fig. 81). The result is a substantial quantity of eroded and eroding earthen architecture across the monumental complex.

At Merv much of the unexcavated archaeological site, alongside the eroded defences, has been left untouched. Today the eroded and eroding defences pose problems as they are undercut at the base, and suffering from water erosion at the wall top, this is problematic as the upstanding elements of the defensive walls give shape and form to the archaeological site and complex of cities. In contrast at Shahrisabz (Uzbekistan) stretches of the city wall have been left untouched, and these remain extant in an eroded and eroding form, used as property boundaries and surviving in stark contrast to those areas of the city wall that have been conserved through encapsulation (see below).

The various excavated sites at Samarkand and Khiva (Uzbekistan) and Jeitun and Merv (Turkmenistan) all have open and eroding archaeological trenches. Leaving the archaeological trenches open is associated with the nature of archaeological work within the Soviet system (Masson 1989; see above backfilling). However in the long-term 'doing nothing' has resulted in substantial erosion and damage to the excavated and surrounding unexcavated archaeological deposits.

'Do nothing' assessment

Practical impacts

In the long-term the decision to 'do nothing' will dictate the survival or otherwise of a site, and will result in the continued and continual erosion of the structure or site. Without measures taken to correct and retain the structure or site in the present form the continued erosion of the abandoned structures will result in the more complete collapse and erosion through time. For those structures retaining protective roofs, and internal decorated features and niches, the continued loss and erosion will result in substantial loss to the surviving fabric when the roofs collapse. The integrity of these structures is therefore placed at considerable risk through the continued erosion and eventual collapse of the protecting roofs.

On excavated archaeological sites, such as Samarkand and Khiva (Uzbekistan) and Jeitun and Merv (Turkmenistan), abandoning the site with open and eroding trenches has made an enormous impact to both the excavated and unexcavated materials. The open trenches act as drainage sumps and sponges, and further damage the unexcavated strata adjacent to the excavated materials. After a period of time trenches left open and eroding will establish a new equilibrium with the surroundings and the rate of erosion generally stabilises. However in establishing this equilibrium they cause considerable damage to the excavated material both within the trench and in an area that exceeds the area of the trench.

Similarly on archaeological sites (either excavated or largely unexcavated), such as Merv, Çatalhöyük (Turkey) and Afrasiab (Uzbekistan) the tell has been left in its eroded and eroding condition. Damage is therefore occurring to the unexcavated materials retained within the tell, this is a result of wind erosion and tell creep, water erosion, and in other instances there are problems associated with vegetation growth and damage as a result of burrowing animals. This is particularly problematic as conservation activities tend to be undertaken on a limited basis, concerned with an intervention or area rather than considering the whole site.

In contrast if 'doing nothing' occurs in a context in which sites are not threatened by other factors (such as development, quarrying or looting) the site may be retained in its eroded and eroding form for an indefinite, but lengthy period.

Current conservation theory

Conservation theory is characterised by advocating 'doing something' and carrying out some sort of intervention on the historical or archaeological fabric. The 1999 Burra Charter states: "do as much as necessary to care for the place and to make it useable, but otherwise change it as little as possible so that its cultural significance is retained." (Burra Charter Preamble). As such 'doing nothing' is not an approach present within conservation charters.

Conservation theory is applicable to sites, monuments and buildings that are already valued and have a high profile. As a result a number of sites in which a 'do nothing' approach is recorded are sites that are low profile and therefore there is no incentive to place the monuments and sites within proactive conservation and management contexts. The process by which value is assigned to buildings and sites is complex, and may result in different responses, one of which is retention almost always achieved through some sort of conservation and management intervention.

In other instances people are aware of the site, but the act of remembering and valorising the site manifests in different ways to proactive conservation and management. On a number of sites within the study area (Anau (Turkmenistan) and Nurata (Uzbekistan)) parts of the monuments are valued for the symbolic and ritual associations whilst other parts of the site are left alone and visited as part of ritual and symbolic activity, and so the Mosque complex at Anau is actively conserved, whilst other parts of the site suffer from continued erosion. Conservation theory would advocate a more holistic approach to the conservation and management of the whole site, its various values and cultural significance (Burra Charter Article 1). Such an approach would place the site in a conservation and management context in which funding, and political will was such to assure the survival of the sites cultural significance, assuring compatible use and would take into consideration measures that could be taken to retain and protect the sites setting (Burra Charter Article 8). In the case of Anua and Nurata, this would be concerned with retaining contemporary use, alongside the conservation of the tell.

On archaeological sites around the world there is a problem with the abandoning of trenches after excavation. In the study area the leaving open of excavated trenches after investigation was associated with the political and ideological context in which archaeological research was carried out within the former Soviet Union. Within this context the leaving open of excavation trenches would enable the 'past' to be seen and interpreted by those who visited a site. The legacy of this approach is the thousands of open, and now abandoned and eroded archaeological trenches on sites throughout the former Soviet Union. Conservation charters, such as the ICOMOS Charter for the Protection and Management of the Archaeological Heritage 1990, would advocate funding and provision be made for the conservation and management of excavated trenches, through the provision of shelters or through the selective backfilling of zones not left open for 'viewing'.

The application of conservation theory is similarly affected by the profile and value attached to a site or monument. This determines the financial, economic and political capacity for conservation and management activities. Therefore those sites with a low profile are less likely to attract funding and much more likely to have 'nothing done' (or only limited actions) to promote management and conservation.

In locations in Shahrisabz (Uzbekistan) where a 'do-nothing' approach has been adopted for the city walls, these walls now form the property boundaries, in effect when these eroded and eroding sections of wall have gone the position will be imprinted and retained within the town plan. Though in a poor condition, the defensive walls in Shahrisabz retain much more significance than the walls conserved within the other Uzbek cities of Khiva and Bukhara. In Shahrisabz there is the sense that the value associated with the urban setting has been retained, and again this reflects contemporary conservation theory.

Values of earthen architecture

'Doing nothing' both challenges and re-inforces the positive and negative values associated with earthen architecture. The sheer scale and number of historic and archaeological sites left eroded and eroding around the world attest to the universality of earth used in the construction. On a number of sites where 'doing nothing' has been an option, the degree of preservation and retention of earthen architecture is a testament to the durability of earth used in construction. In the case of Nurata (Uzbekistan) the earthen walls have been left eroding and the current condition is testament to the ability of the originally massive earthen architecture to retain its shape and form regardless of long-term deterioration. In addition the type and extent of earthen architecture attests to the great local and temporal distinctiveness in the materials and techniques utilised.

'Doing nothing' to earthen building materials allows the materials to erode, and be further eroded through time. On the one hand, this process of deterioration, formation and deformation enables earthen building materials to be re-used and recycled, and sites and structures to change shape, form and use through time. This may be a positive attribute of earthen architecture; but in other instances, this reinforces the negative associations of earthen architecture as an 'unconservable' building material.

'Doing nothing' presents us with a considerable paradox: this is particularly so as we tend to view sites on an intervention by intervention basis, rather than viewing the whole site. As a result we tend to see individual trenches and buildings retained in a poor condition (reinforcing the negative associations of earthen architecture) whilst ignoring the bigger picture of the retention of the whole site (showing the positive associations and longevity of earthen architecture)

Sustainability

'Doing nothing' on a historic or archaeological site affects the type, nature and form of legacy handed on to future generations. This conditions the understanding and interpretation of the archaeological and historic environment.

'Doing nothing' may be an appropriate and sustainable solution if it is accompanied by other types of intervention such as documentation, and the placing of the historic or archaeological site within a management context that reduces the risk of rapid loss associated with development, and quarrying (see discussion *Chapter 17*).

However this approach is particularly problematic if 'doing nothing' becomes a response to a site or building after some intervention has already taken place. On archaeological sites the failure to provide for the conservation of the excavated materials poses threats not just to the excavated material but also to the adjacent unexcavated material. In the long-term the limited conservation threatens the very values that make archaeological sites important. On archaeological sites it is unusual, although not uncommon in the 21st century to come across such an unsustainable approach to excavation, conservation and management.

Similarly emblematic of a lack of sustainability in the approaches taken for the retention and conservation of abandoned structures comprising earthen architecture is the abandoned restoration work that was being carried on the large *qala* in Shahdad (Iran). It is difficult to plan for the adaptive re-use of the large abandoned structures, given the costs implicit within restoration, and the difficulties of adapting these structures to fulfil a role within modern society. This is problematic in Shahdad due to the number of abandoned structures and the remote location. The modern village has also been subject to more recent changes associated with the decline in the rural population. The resulting mass of abandoned structures poses problems associated with maintenance and/or retention, in which 'doing nothing' is the only economically, politically and practically feasible option.

As buildings erode into archaeological sites and archaeological sites continue to be subject to agencies of formation and deformation, future generations may have only the archaeological earthen materials rather than the upstanding building or structure upon which to understand and interpret the past environment. There is a formidable tension between 'doing nothing' to a site, and allowing deterioration, formation and deformation of earthen archaeological deposits (and possible recycling and re-use), and the notion of equality and sustainability between generations. Is it right now to limit the potential use and re-use of buildings, monuments and sites by future generations? This paradox is also concerned with what is and what is not valued, and by being valued conserved. If it is the historic fabric and form, then clearly 'doing nothing' is not a sustainable approach; but if it is the values (and changing values) associated with a place than 'doing nothing' does not impact the future sustainability.

Chapter 12: Drainage and undercut repair

Drainage comprises the measures taken to direct or re-direct falling and rising water away from earthen architecture, these may be concerned with the protection of a single wall, entire monument or site. Drainage is concerned with the redirecting of water to prevent the formation of drainage gullies and/or erosion to the wall base, and involves alteration to the monument or site setting, through the re-directing of water away from the main body of the archaeological or historical fabric. Undercut repairs are the repair, and reworking of the zone at the wall base most prone to, and affected by damage from rising water (undercutting). Undercut repairs are carried out on both upstanding earthen structures, and on the excavated walls and section baulks uncovered through the course of archaeological excavation.

Documented research has been concerned with developing methodologies for the installation of suitable below ground drains, investigating issues such as separation (for reversibility), choice of drainage materials and monitoring. Much of the work on the application of drainage solutions to historic and archaeological earthen sites has borrowed from methodologies developed for ground engineering applications and other related disciplines (such as agricultural drainage). The available documentation indicates some research has been carried out on earthen archaeological sites in the USA, Syria, and Pakistan. Within the site dossiers different drainage solutions were recorded in the study area in Turkmenistan at Merv, Nisa and Gonur. In contrast the methods and materials for undercut repairs are poorly represented in the documentation. Although they do represent one of the earliest conservation approaches for earthen architecture utilised at Casa Grande Ruins (USA) at the turn of the 20th century (Matero 1999; Matero *et al* 2000). The documented research has developed methodologies for repairing wall bases, identified the nature of materials used, and appropriate methodologies for cutting out and infilling the damaged zone. This research has been carried out on sites in Peru, USA and Syria. Within the site dossiers undercut repairs are recorded in the study area at Gonur and Merv (Table 9 & 10).

A variety of different approaches have been utilised for drainage and undercut repairs, using a variety of different materials and techniques.

Below ground drainage
Below ground drainage work involves measures that will assist in discharging falling water (and limiting rising water) from part, or all, of the monument or site setting. Often this is concerned with drains intended to encourage water to drain away from wall and section bases in order to prevent erosion associated with rising water and undercutting at the wall base. French drains (and soakaways) are small trenches, dug to a gradient to allow surface water to drain away from a building or area at risk. The trench is filled with gravel or aggregate (often with a plastic land-drain placed at the base of the trench), sometimes utilising geotextile fabric to prevent the drain becoming clogged up with fines washed out of the soil. These trenches run into an additional soakaway to further disperse water.

Drainage from the historic or archaeological fabric
Drainage from the historic or archaeological fabric involves measures that will direct water away from the main body of the structure. This may be through the repair of existing drainage and down pipes, or through the installation and formation of new drainage and down pipes. As with capping, drainage is concerned with directing water away from the main body of the fabric, and preventing the formation of vertical drainage gullies. The materials utilised for this sort of work are often harder, replacement materials such as fired brick, tile or concrete/cement for lining the drain (especially on down pipes), or the installation of ceramic, metal or plastic drainage pipes intended to collect water and allow it to be directed away from the main body of the wall.

Other preventative drainage measures
Other preventative drainage measures may be concerned with altering the pattern of water movement by preventing water from collecting in locations that may erode or further erode vertical gullies through the historic or archaeological fabric. Simple measures such as plugging the top of drainage gullies at the first sign of formation, and annually thereafter, can assist in preventing the formation, and worsening of vertical drainage gullies. In other instances preventative drainage measures comprise the building of low mudbrick walls to re-direct water run-off from further eroding undercuts and wall bases.

Undercut repairs
As with many of the conservation approaches these utilise either replacement earthen materials or replacement harder materials, such as cement and fired brick. The design of the undercut repair determines the nature of work undertaken below the ground (as some utilise drainage works), and nature of join and tie-in with the original, historic fabric.

Sites
At Merv, a section of the medieval defensive wall had below-surface drains installed as part of experimental work in 2002. These belowground drains were constructed by lowering a small rectangle and/or diagonal drainage gully through the ground surface and filling in the empty zone with rounded river pebbles, and geotextile (Fig. 135). Other preventative drainage measures at Merv have been concerned with the protection of the upstanding monuments through the addition of materials to the structures surfaces to prevent water from collecting and causing further erosion. In other locations mudbricks have been used to create low protective walls to prevent water run-off into particularly problematic zones, at the wall bases in the Great and Little Kyz Kala and the Palace in Shahriyar Ark, with additional low walls constructed to prevent water from draining into the exposed lower-storey rooms on the Little Kyz Kala (Fig. 137). Additional preventative drainage measures have

also used small earth plaster plugs to fill-in and block the top of drainage gullies to prevent water run-off from entering the already eroded gully and further eroding the vertical faces of the structures (Fig. 134). This sort of work carried out on the Little Kyz Kala, and excavated section through the Gyuar Kala defensive walls requires annual maintenance (Peek 2004).

Similarly a variety of different materials and techniques have been used for repairs to undercut wall bases at Merv (*Appendix 6*). A number of different methods have been used on the Great and Little Kyz Kala, and Kepter Khana buildings. The different methods have been used to identify the best and most effective method of assuring connection between the old and new work, and establishing the original shape of the eroded wall bases. For example, the work on the Great Kyz Kala utilised mudbrick, earth mortars and an earthen render on the surface (although not on the section), on top of this earth plaster has been shaped to form the junction between the buttress repair and the eroded base of the corrugation (Fig. 154-155). Similar work was carried out on a stretch of the Sultan Kala work used in association with drainage and encapsulation interventions (Fig. 145).

Work on the Kepter Khana at Merv has used fired brick and mudbrick, sometimes with and sometimes without, separation materials such as geotextile (*Appendix 6*; Fig. 142, 144). On the Kepter Khana the work has been carried out to use the materials inserted into the undercut to reconstruct the original angle and dimensions. In other instances heavily undercut walls have been supported by the erection of supporting buttresses, constructed from a combination of fired brick and mudbrick.

At Gonur (Turkmenistan) a number of experiments were undertaken to test for the most effective method for repairing the damaged wall bases of the excavated material within the palace complex (*Appendix 6;* Fig. 143). This testing has utilised a variety of different materials and techniques. Including modern fired brick and cement mortars (used in combination with modern mudbrick and earthen mortars). In a number of instances the undercut repairs have been separated from the historic fabric with the placement of geotextiles, within the undercut zone.

At Nisa (Turkmenistan), surface and below surface drains have been installed in the bottom of the excavated complexes and rooms (Fig. 138). The drains are constructed by lowering a small rectangle of the ground and filling in the empty zone with aggregate. On other sites such as Aztec (USA) and Chaco (USA), surface drains have similarly been installed as part of the backfilling designs that allow water to drain through the bulk fill, or allow water to be collected (Fig. 139-141).

In other instances small earth plaster plugs have been used to fill in drainage gullies to prevent water run-off further eroding the vertical faces of the excavated structures. Similar work was also carried out at Gonur (Turkmenistan) in 2004, where vertical gullies were installed down the upstanding excavated walls; these gullies were constructed utilising harder materials, such as cement.

Drainage and undercut repair assessment

Practical impacts

Of the interventions carried out on earthen architecture, drainage and undercut repair can work particularly well. Drainage work protects upstanding and excavated earthen walls from the most damaging effects of rising and falling water, and protects wall bases from further deterioration and undercutting. For a monument or structure already suffering damage to wall bases, undercut repairs can limit further damage associated with rising water, particularly so as continued erosion damages the inserted replacement materials rather than the historic or archaeological fabric. Similarly some of these undercut repairs, (for example those used at Merv) are utilised alongside invasive drainage solutions that result in substantial alteration and damage to the surrounding unexcavated archaeological deposits (see drainage above).

Once the area to be protected by the installation is assessed the use of drainage and undercut repair is effective, particularly when used together. If they are well maintained they can successfully re-direct water away from the wall and/or site and therefore remove damage and deterioration factors such as falling and rising water from the structure and/or site.

Some of the drainage and undercut repair work can alter the shape and form of the exposed historic or archaeological fabric. The visual impact can make interpretation of the excavated complex and/or building complicated, and this is particularly the case when materials are added to the site and/or building, such with the use of gravels in surface drains and the utilisation of more complex geotechnical drainage solutions. The installation of new material at the wall base alters the sites setting and visual characteristics. The visual impact can make interpretation of the excavated complex and/or building complicated, and this is particularly the case when 'foreign' materials are added to the site and/or building. The visual impact is greater when it may be desirable to 'recreate' the wall base in a shape and form in which it appeared in the original. This is problematic as the current shape and form of a structure is already eroded, and substantially reduced from the 'original'. As a result tying-in the reconstructed line can be problematic, resulting in a substantial alteration to the site or building shape and form. Such is the case with the undercut repair undertaken on the north-western corner of the Kepter Khana at Merv (further discussed below).

Some of the solutions may be invasive, such as excavating below the ground surface for the installation of drains/soakaways, and the removal of the already damaged zone at the wall base. In all instances this involves damage to, and removal of, unexcavated archaeological material. This is problematic as often conservators carry out this sort of work rather than archaeologists and the excavation may result in the undocumented destruction of archaeological deposits.

Sometimes these solutions work too well and reduce the moisture content of the surrounding zone. This may damage the surviving historical and archaeological fabric, through desiccation and/or the alteration of natural run-off patterns. This may reduce the chances of survival of the structure being conserved, alongside altering the characteristics of the below-ground buried archaeology.

One of the problems most associated with the installation of undercut repairs is the lack of tie-in with the original historic fabric, and problems associated with the drying, shrinkage, cracking and separation of earthen materials used in repair. As a result a number of different materials have been experimented with including the use of ceramic or metal rods inserted into the historic fabric to tie in the new work, and the use of brick inserts to infill cracks that may appear at the interface between the inserted and historic materials.

As with most other conservation work undercut repairs rely on regular maintenance and re-working to assure effectiveness. If this is not carried out drains may become clogged up or repaired zones can result in substantial damage. The requirement for maintenance is particularly acute for earth plugs as these are washed out and eroded by rainfall.

Some of the materials and techniques utilised for undercut repairs can contribute to increased erosion. For example, the experimental approaches recorded in 2004 at Gonur (Turkmenistan) were concerned with the utilisation of harder elements such as concrete and fired brick. The problems associated with the use of these materials are noted above.

Current conservation theory

As specific techniques, drainage and undercut repairs are not advocated within conservation charters in the same way as other techniques, such as backfilling and sheltering. However drainage and undercut repairs do fall into a broad category of maintenance activities that may be appropriate for earthen architecture, and this category of intervention is mentioned within conservation charters (Burra Charter Article 16).

Drainage and undercut repairs may also cause alteration to a site or monument setting, as such some of the concerns within conservation charters may be appropriate to consider. For example, the Burra Charter states:
"Conservation requires the retention of an appropriate visual setting and other relationships that contribute to the cultural significance of the place.

New construction, demolition, intrusions or other changes which would adversely affect the setting or relationships are not appropriate." (Article 8).

Although this is not particularly explicit this does emphasise the need for these types of work to be planned appropriately in order to prevent damage to the monument or site setting. This is particularly important on archaeological sites where drainage and undercut works may destroy or damage unexcavated deposits.

The use of these solutions meets the requirements of contemporary conservation theory by being concerned with holistic approaches for the retention of the archaeological and historic fabric (such as advocated through the Burra Charter). By looking at and remedying the wider effects of erosion and deterioration both drainage and undercut repairs differ significantly from piecemeal interventions on the historic or archaeological fabric. These types of interventions also fulfil the requirements of conservation theory by retaining the historic or archaeological fabric with only a minimum of intervention. Undercut repairs can avoid more substantial intervention on the archaeological and historic fabric, such as reconstruction and encapsulation. The type of repair, and the materials used impact other notions particularly as replacement materials impact the authenticity and integrity of the 'original' fabric.

The installation of drains and the repair of undercuts are often irreversible as the process of excavating below the ground removes deposits, and the materials used may be difficult to remove. In other instances the use of earthen materials for the plugging of the top of drainage gullies means the earthen material is washed out, deposited and redeposited over the face of the surface. These all substantially alter the form and appearance of a site or structure, which is not necessarily advocated through conservation theory (for example the Burra Charter states conservation approaches should not alter or impact the cultural significance of a place (Article 1.1)).

Most of the drainage and undercut work is clearly distinguishable from the historic or archaeological fabric. In a number of instances problems are associated with the appearance of some of the work (particularly the gravel rectangles, and surface drains) and these have the potential to be misunderstood and interpreted as unusual archaeological features. As a result some of the techniques and materials can impact the values associated with an archaeological site.

Symptomatic of the tensions posed in relation to conservation theory are problems associated with the tie-in and line with which work should be carried out. This blurs the distinction between undercut repairs as minimum interventions and more substantial intervention (and partial restoration). This is illustrated by the problems posed by the undercut repairs on the north side of the Kepter Khana at Merv. Here, the erosion of the structure has impacted the archaeological understanding and interpretation of the original shape and form. Therefore the undercut repair could be at the eroded line and form, or at the hypothesised original line (extending the current dimensions for the upper, better preserved parts of the structure). The various different solutions adopted here (with the north-east corner rebuilt at the eroded line, whilst the north west corner is rebuilt at the proposed original wall line) results in confusion, and illustrates how relatively simple interventions such as undercut repairs rely upon the documentation and understanding of the monument, alongside the assessment of the different approaches in relation to conservation theory.

Both approaches present tensions by threatening the unexcavated archaeology through the need to excavate and remove damaged wall bases and install drains. Of all of the conservation techniques utilised for earthen architecture, drainage (and some methods of undercutting repair) poses the greatest problems in terms of archaeological impact, and so the work needs to be archaeologically recorded prior to the commencement of conservation work. A value is placed on the upstanding historic or archaeological fabric in preference to that which is retained below the ground. If conservation is concerned with the retention and management of all of a sites cultural significance (Burra Charter Article 1), then these types of work need to be planned within wider and holistic conservation and management planning for the whole site.

Values of earthen architecture

Drainage work and undercut repair require the identification and understanding of the factors resulting in damage - these factors are often environmental. As such these 'environmental' solutions are suitable for a material that reflects positive environmental qualities.

Both of these approaches mimic the type and nature of interventions undertaken on earthen architecture in living contexts (such as the installation of damp-proof courses). As such drainage work reflects the type of maintenance and repair that occurs to earthen buildings in contexts of use. Therefore this is an approach to the conservation and management of earthen architecture that reflects the values of earthen architecture as a material that can be damaged by the environment but is also made durable through appropriate adaptation.

Some drainage and undercut work changes the visual component of earthen architecture. For example drainage utilising non-earthen elements changes the visual and aesthetic component of sites, this is particularly so when vertical drains are installed (as at Gonur (Turkmenistan)) and where gravel is placed on surface and french drains. This impacts the aesthetic values associated with earthen architecture, changing its soft, continuous contours to more rigid and solid lines.

In addition the work may reduce the local distinctiveness of earthen architecture (repairing an undercut on a placed earth wall with mudbricks), or alter the construction type (repairing and undercut mudbrick wall with fired bricks or concrete), as this new work tends to dominate sites some of the drainage work has the tendency to make sites look the same and results in a loss of local distinctiveness. As with all of the conservation solutions, the use of replacement materials implicitly suggests something is 'wrong' or 'negative' with the original.

Sustainability

These solutions can work well and assist the longevity of earthen architecture, assuring the legacy is retained by the present generation for future generations. By looking at the causes of deterioration and erosion and trying to limit the effects in a wider area many of the drainage and undercut repair solutions are associated with more holistic and sustainable approaches to site conservation and management when compared with other for individual walls (such as consolidation or encapsulation).

Drainage and undercut repairs rely on monitoring and maintenance to ensure they remain effective, for example checking any shrinkage that may become apparent on drying. As with many of the solutions for earthen architecture, there is a tendency to see these interventions as a 'one-off' solution after which there may be a decline in interest in the work carried out, this is problematic as all conservation work requires maintenance. In the long-term if commitment to monitoring and maintenance is not retained these approaches may result in substantial damage, threatening the sustainability of the remaining historic and archaeological fabric. Therefore the sustainability of these actions is only assured if finances are made available to permit these activities subsequent to the initial work.

The sustainability of some of the drainage and undercut repairs are questioned as by excavating material away from the base of wall and/or archaeological sites, a value is placed on the material in its present shape and form, rather than on the unexcavated archaeology, this implies that the unexcavated archaeology can be 'sacrificed' to make way for conservation interventions to retain the extant wall line.

These solutions are very effective as a means to change local moisture regimes, and so in those areas threatened by wider climate change care needs to be exercised when planning for and implementing drainage schemes, in order to incorporate within the design solutions that may accommodate future changing environmental patterns (such as increased waterfall, fluctuating water tables, and desertification). As such some drainage solutions may need to be at a larger scale to manage increased run-off, whilst in other locations threatened by desertification, it may be unnecessary to install drains, as water dispersal may acelerate aridity and actually result in an increase in loss due to the shrinkage and cracking of the clay component of earthen architecture.

Chapter 13: Maintenance

Maintenance encompasses all of the activities associated with prolonging and keeping a monument or site through preventing deterioration, such as re-mortaring and replastering. Maintenance is often associated with the prolonging of the life of earthen structures in living contexts (*Chapter 6*), but is also used as a conservation and management tool for buildings and archaeological sites.

The pattern of maintenance is concerned with the annual maintenance of roofs, through to the reapplication of earthen renders and plaster, alongside other interventions and repair such as infill of damaged patches, repairing damage at wall bases, and repair of 'at risk' areas such as drains. The materials used for maintenance are variable, altering from the use of traditional earthen mortars and renders, to the use of replacement materials, such as harder, cement-based materials for re-mortaring and re-plastering.

As an approach to the conservation and management of earthen architecture maintenance is not particularly well represented within the documented research (other than within anthropological and ethnographic accounts in 'living contexts' (*Chapter 6*)). Within the study area, and site dossiers maintenance activities were recorded within the residential parts of Khiva and Bukhara (Uzbekistan), Yazd (Iran), Konya (Turkey), and more unusually in Turkmenistan at Taklahtan Baba and Merv (see 169; Table 11).

Earthen mortars and renders

In the residential zone of Ichin Kala in Khiva (Uzbekistan), structures exist within contexts of maintenance and repair (*Appendix 6*). The types of maintenance in these residential areas are primarily concerned with the annual maintenance of roofs through the reapplication of earthen renders and plaster, and the annual maintenance of walls with an earth plaster. Other interventions and repairs of the earthen buildings in contexts of use and maintenance include the infill of damaged patches, repairing damage at wall bases, and repair of 'at risk' areas such as drains.

In the residential zone of Bukhara (Uzbekistan), a variety of different earth building traditions can be found, primarily *sintch* and *paksha* alongside mud mortars, mud plasters and mud renders (*Appendix 6*). As is appropriate for earthen construction these structures exist within contexts of maintenance and care (Fig. 29, 30). The materials used for the maintenance and repair of the domestic buildings are a hybrid mix of traditional earthen plasters, renders and mortars alongside other modern imported materials such as cement and fired brick.

In Yazd (Iran) the historic town is being repopulated with a swelling refugee population. The economic capacity of the refugee populations means that the structures they re-occupy are retained and maintained using cheap and available materials and techniques, such as mudbricks, earthen mortars and earthen renders. These are in contrast to the approaches used by the more affluent members of society utilising materials such as concrete and fired brick (*Appendix 6*).

Replacement materials

In Konya (Turkey) a number of the domestic buildings were in a very good state of maintenance, whilst others had suffered from a lack of maintenance to the surface plasters (*Appendix 6*). The most common exterior render seen in Konya other than the mud-straw plasters was a yellow/orange and green paint applied on top of a thin cement render (in other instances the cement render was left unpainted). A number of these buildings have suffered significant damage as a result of these cement renders and significant damage has occurred to the exposed earthen wall fabric. In many instances substantial cracking leading to the detachment of the cement render causes problems. Patched repair jobs normally using cement were present on many of these buildings, and as infill where doors and windows had latterly been inserted. Similar utilisation of cement renders for plastering was recorded in Yazd (Iran) in those zones in which more affluent members of society utilise cement-based materials for the maintenance of residential and restored structures (*Appendix 6*).

In the residential zone of Bukhara (Uzbekistan), one of the repairs to a *sintch* house used fired bricks as infill (*Appendix 6*). It was uncertain if the domestic structure had remained occupied during the Soviet period or if these were structures that had been re-occupied post-independence, and therefore reflected changing contexts of use of building materials and techniques. The re-occupation after a period of abandonment would therefore have resulted in more substantial repair needs, this alongside the changing nature of traditional skills and techniques, probably accounts for the hybrid variety of materials and techniques used in the residential zone of Bukhara.

Other

In Turkmenistan at Taklahtan Baba and Merv a very different use of earth is represented in Turkmen funerary rites (*Appendix 6*). Where burial occurs beneath a low earthen mound (Blackwell 2001), the burial mounds are maintained with the annual re-application of earthen plaster (*pers comm.* Gaigysyz Joraev). In addition a number of more recent burials have utilised fired brick for the construction of low walls and cement for the capping/plastering of the burial mound. The shrine buildings associated with these complexes are also maintained through the re-application of earthen plasters and renders. It is interesting that in Turkmenistan (as elsewhere in Central Asia) that burial and burial rites are places that are re-visited and maintained with the same care and regular maintenance as homes, as such they are 'living places' (see *Chapter 6*).

Maintenance assessment

Practical impacts
The practical impact of maintenance is the retention of earthen architecture for the life of the building or duration of the sites occupation, and maintenance can assist in the rejuvenation of interest in traditional building materials and techniques.

The impact of changing maintenance regimes and the utilisation of replacement materials for plaster and render, is such that they can accelerate deterioration and erosion. In Bukhara (Uzbekistan) and Konya (Turkey) the use of cement-based materials has created a hard impermeable layer, below which is an area of increased erosion. In Konya (Turkey), where the cement renders on the domestic buildings have fallen away, significant damage to the wall core is visible. Damage to the wall bases was noted as a result of splash back from vehicle access, alongside the increased humidity from loose, stacked earth and rubbish at the wall base. This sort of damage is exaggerated where renders are continued and extended to the wall base, effectively making the damp-proof coursing and capillary breaks redundant.

The visual impact of maintenance is to create an appearance of 'newness' through the use of earth plasters and renders (for example within the Arg-e Rayen (Iran), Fig. 108). This enables a site or building to be understood and interpreted in its current form. However the impact of maintenance activities on archaeological sites (often as part of encapsulation work – see above) can add some confusion to the understanding and interpretation of a structure, as it is difficult to understand the 'new' appearance, particularly when this has covered over the phasing or construction details.

Stressing the importance of maintenance Balderrama and Chiari (1995) comment:

"It should be stressed once again that the key point in the conservation if a fragile materials like mudbrick has always been maintenance. Without maintenance there is no hope of preserving monuments in mudbrick, whatever treatment is performed. On the other hand, good, careful maintenance may sometimes give better results than the most sophisticated and expensive treatments." (Balderrama and Chiari 1995, 106).

Current conservation theory
As has already been seen, continuing occupation of structures is identified as an important conservation approach (*Chapter 6*). Similarly the Venice Charter stresses the importance of maintenance in living contexts, and also maintenance of conservation interventions:

"It is essential to the conservation of monuments that they be maintained on a permanent basis." (Article 4).

The Burra Charter provides greater clarity in defining what 'maintenance' constitutes:

"Maintenance means the continuous protective care of the fabric and setting of a place, and is distinguished from repair. Repair involves restoration or reconstruction." (Article 1.5).

The explanatory notes then make distinction by providing the example of roof gutters, where maintenance would be regular inspection and cleaning, whilst repair involving restoration would be the returning of dislodged gutters, and repair involving reconstruction would constitute replacing decayed gutters (Burra Charter explanatory notes to Article 1.5). However the types of maintenance advocated through conservation charters tend to be those concerned with the maintaining and checking of conservation works. For example the ICOMOS Charter for the Protection and Management of the Archaeological Heritage 1990, recognises the pragmatics and difficulties of maintenance of conservation work stating:

"Owing to the inevitable limitations of available resources, active maintenance will have to be carried out on a selective basis" (Article 6).

This is a very different type of maintenance than the regular activities associated with earth structures such as replastering.

The use of maintenance is complicated as older approaches to conservation theory advocate that value rests with the age and visibility of the archaeological and historical fabric. For example the SPAB manifesto comments:

"...the whole surface of the building is necessarily tampered with; so that the appearance of antiquity is taken away from such old parts of the fabric as are left, and there is no laying to rest in the spectator the suspicion of what may have been lost; and in short, a feeble and lifeless forgery is the final result of all the wasted labour." SPAB Manifesto

As a result maintenance of exposed surfaces utilising earthen plasters and renders has been avoided (often in preference for experimental work concerned with consolidants that would retain the visibility of the surface of earthen architecture - see above).

Whilst in many respects the current (post Nara) period of conservation theory has moved significantly beyond the SPAB notions, the issues of visibility and the appearance of 'newness' remain complicated, particularly in different contexts. For structures that remain in residential use, such as those in Uzbekistan at Bukhara and Khiva, the maintenance of the mudbrick and *sintch* structures relies on renewal of earthen renders and plasters on walls and roofs, giving the appearance of 'newness' regardless of the age of the structure. Whilst this is an appropriate response to the maintenance needs of earthen architecture in living contexts, this approach may not necessarily be appropriate to earthen architecture in archaeological or abandoned contexts.

The utilisation of earthen renders and plasters has been perceived as 'obliterating' the archaeological or historical fabric, and altering the appearance of a site or building to the extent that the age value is lost. These ideas have added complications to the assessment of different approaches for the conservation of earthen architecture in light of conservation theory. For example the Çatalhöyük (Turkey) management plan states:

"Adobe construction traditionally depends on ongoing maintenance procedures. In the Konya region typically a mud slurry is applied to the external surfaces of adobe buildings every few years. The application of new surfaces to ancient materials or surfaces, however, obliterates their conservation." (Çatalhöyük Management Plan 23)

In some instances this notion has some grounds for support, such as when a conservation solution is sought for excavated archaeological sections (where it is the visibility of the phasing that is sought for the interpretation and understanding of the whole site), or on sites and buildings retaining earthen architecture in a form in which it was never coated in an exterior plaster (such as placed and rammed earth walls), or, as at Çatalhöyük, where conservation is sought for excavated and exposed decorative details and wall paintings.

The utilisation of harder, cement based materials for maintenance poses problems as they can accelerate the erosion associated with earthen architecture and pose problems in relation to the notions of authenticity. In other instances the use of earthen renders and plasters for maintenance is an appropriate conservation solution. Conservation theory would advocate that the materials and techniques utilised for conservation should be assessed for the wider impact, and planned for within contemporary approaches to conservation and management planning. If conservation theory is concerned with the retention of all of the values associated with an archaeological or historical site, then one of those values is the renewal of earthen architecture through maintenance.

In the residential areas of Khiva (Uzbekistan) earthen architecture is retained in contexts of use and maintenance, and the relationship with the structures is active, and suitable given the values and needs of earthen architecture. The dichotomy seen in historic sites such as Khiva in conservation approaches to the monumental and the residential structures is a legacy of the tension between public and private ownership, and also indicative of the types of structures that were assessed as worthy for conservation and restoration in the past. Conservation theory would advocate a more holistic approach to the conservation and management of the historic town, stressing the importance of the monumental and domestic structures, alongside the communities and people who inhabit them. Similarly in Yazd (Iran) there is a great distinction between the new residential suburbs and the historic town. The declining population within the old town has resulted in the retention, and current very active conservation, of the higher status buildings (see restoration below). In these contexts it is conservation without people that has resulted in the use of harder, cement based materials to reduce the requirement for maintenance and replastering. In this instance conservation theory would perhaps seek a greater interaction between the conservation approaches and contemporary society.

Values of earthen architecture

Maintenance is associated with humanity and locality and is a significant and positive attribute of earthen architecture. The very active maintenance of the residential buildings in Bukhara reflects the perception of earthen architecture as durable, living and breathing. This contrasts with the static monumental and defensive structures in Bukhara retained through the use of replacement materials.

The work in Yazd (Iran) is important in showing pragmatic approaches to the retention and re-use of earthen architecture. In the re-occupied residential areas two approaches are adopted, the wealthier parts of contemporary society using the models of the 'new life for old structures' campaign - using coloured cements and non-earthen materials for reconstruction whilst retaining the appearance of the 'old' (see restoration below); and for the refugee communities, the use of available earthen building materials and techniques. This dichotomy of approaches indicates the complex, social, economic and political context of conservation and rehabilitation of the old town in Yazd. Both approaches change the values of the earthen architecture retained in Yazd, making earth structures durable, but 'cheap'; and in contrast retaining the aesthetic values of earth but re-interpreted through the use of non-earthen materials.

In Konya (Turkey) the use of cement renders on the domestic earthen architecture is indicative of changing patterns of traditional building skills and materials. Change is associated with wider economic, social and cultural change and the negative perception of earthen architecture. The purpose of these cement renders has been to add perceived qualities of longevity (and reduced maintenance) to the buildings. They also change the appearance of buildings to 'appear' more long lasting and durable. However as the example of Konya shows, the use of cement materials accelerates damage, further reducing the longevity of the earthen structures, and influencing further the negative associations of earthen architecture by re-enforcing perceptions of weakness.

Sustainability

Maintenance regimes make reference to the activities carried out in past, which are utilised today, enabling structures and sites to be retained in the future. The sustainability of the approach is unquestioned, for regular maintenance can prolong the life of structures indefinitely. Maintenance also relies on the engagement and employment of people to carry out the work. This can increase the sustainability of the conservation approach through employment and the generation of an economic incentive for conservation.

The most significant challenge to sustainability posed by maintenance is in the choice of materials. With harder cement renders and mortars being ineffective and difficult to remove. This compromises the sustainability of these interventions by limiting the options available for conservation and management in the future. Rectifying the damage that has, and is occurring to the traditional buildings on account of the use of the cement renders is much more problematic, since these properties are retained as dwellings and require complex negotiations and discussions with owners and occupiers.

The major factor affecting the sustainability of the conservation and maintenance activities in many of the towns and cities visited in the study area is the rupture in the life of the old town associated with population movements in the 20th century. There is a danger that the current conservation works have not fully engaged the local population and they will suffer from a lack of maintenance in the long run. This is shown by the use of cements and cement based renders to create the perception of longevity even on the structures that are not wholly occupied. In the long term this approach to the retention of the earthen architecture may prove unsustainable, as damage occurs to the building fabric, and the population is disengaged from the process of conservation.

Today the skills and techniques associated with the construction and maintenance of earthen architecture are affected by the changing social and political contexts in the 20th century. These have altered building practices, and resulted in loss and change to the patterns of acquisition and use of traditional building and maintenance skills.

Chapter 14: Reconstruction and restoration

Reconstruction is the rebuilding of a structure or archaeological site (or part thereof). The purpose of reconstruction is to notionally 'return' a monument or site to a condition in which it was in the past, and/or to an assumed condition in order for a monument or site to regain the appearance that it is interpreted as having in the past. Reconstruction may be *in situ*, or *ex situ* (see removal/relocation below). Restoration is the repair and reinstatement of a monument or site (or part thereof). Often the purpose of restoration is to find and allow new uses for the old buildings. Restoration is more associated with conservation activities for historic, and upstanding earthen architecture, whilst reconstruction is most associated with archaeological sites and abandoned historic structures.

In most instances the materials and techniques used for reconstruction and restoration are alike, utilising either similar, earthen materials or replacement harder materials such as concrete, breezeblock and fired bricks. The work may seek to fully reconstruct a whole building or site, or part of a building or site, often as a didactic tool to assist in interpretation and understanding. In the study area additional zones of reconstruction work have been tacked on to other conservation work, such as encapsulation and capping which has extended to the construction of crenellations on the defensive wall tops utilising either earthen or replacement, harder materials.

Reconstruction is not particularly well represented within the available documentation, with the exception of work in Brazil, China, Cyprus, India, Mexico, Peru and Zimbabwe. Within the site dossiers reconstruction work is documented at, sites in the Southwest USA (Aztec, Bandelier, and Pecos), Çatalhöyük (Turkey), and more unusually the Chapel of Reconciliation in Berlin (Germany). Within the study area reconstruction is recorded in Iran at Rayen and Yazd; Uzbekistan at Khiva, Bukhara and Shahrisabz; and Turkmenistan at Nisa and Merv. The documented research concerned with restoration has occurred in various countries around the world. Within the site dossiers from the study area restoration (or partial restoration of a small part of a monument or site) is documented in Turkmenistan at Anau; Uzbekistan at Khiva and Samarkand; and in Iran at Arg-e Bam and Yazd (Table 12 & 13).

In situ reconstruction - archaeological sites

In situ reconstruction has been utilised in various forms on sites in the USA Southwest, using a variety of different materials and techniques, ranging from the excavated 'original' material, through to earthen materials, and modified and consolidated earthen materials. At Aztec (USA) the site is dominated by the enormous reconstructed central *kiva* (c.50ft diameter, with the floor c. 8ft below ground) (*Appendix 6*). The *kiva* was reconstructed in 1934 and is built using stone removed in the process of excavation, alongside additional stone excavated from further Anasazi villages for the core filler in the restored walls (Lister and Lister 1987). The exterior of the structure was left unrendered (*op cit*). The modern flat roof of the *kiva* has roof drains, and a skylight (incorporated for exhibition and interpretation functions). The massive interior columns were constructed of reinforced concrete and these were subsequently 'hidden' by plaster (*op cit*). Similarly, at Bandelier National Park (USA) a number of the cliff dwellings have been either fully or partially reconstructed (*Appendix 6*). These reconstructions from the early 20th century re-used the original sandstone blocks from the base of the canyon, alongside material that had been excavated from nearby sites. Soil was mixed from the canyon floor for the earthen mortars and interior plasters. In addition some modern materials such as tar paper and newspapers were used within the structures (Rothman 1988). At nearby Pecos (USA) the full *in situ* reconstruction of 2 *kivas* use acrylic-modified earth bricks and plasters. They are presented to the visitor 'as if in use' dressed with ladders and props (*Appendix 6*).

At Rayen (Iran) in those instances where the walls associated with a single structure have been encapsulated in plaster, a small mudbrick and plaster arch has also been reconstructed (see encapsulation and *Appendix 6*). The purpose of the arch is to indicate the shape and form of the original structure, and this helps understand and interpret the plan of the site.

At Nisa (Turkmenistan) one part of the excavated complex has been partially reconstructed, consolidated and maintained, with the use of mudbricks, earthen mortars and renders, and white-wash/paint (*Appendix 6;* Fig. 109). This work has been carried out using a variety of different materials and techniques, primarily building up the missing parts with newly manufactured mudbricks, and the covering of these surfaces with earthen plaster. In this multi-storeyed complex the roofs and ceiling have been reconstructed utilising timbers, reed matting and earthen plasters, alongside cement, and fired brick/tiles applied on top of plastic sheeting. In other instances the excavated plan has been reconstructed with the use of fired bricks to create columns, some of which have been covered in white paint.

At Çatalhöyük (Turkey) a number of elements within 'building 5' were reconstructed following damage to them as a result of water run-off into the excavated area (*Appendix 6;* Fig. 127-128). This reconstruction work utilised earthen materials to recreate the shape and form of the excavated area in reference to excavation drawings and photographs. The surface was then coated in earthen plaster, and consolidated using the same materials and techniques for the unreconstructed parts.

At Merv, the western wall of Abdullah Khan Kala has been reconstructed in reference to historic photographs (*Appendix 6*). The reconstruction work utilised earthen materials (mudbricks, mortars and plasters) to encapsulate the existing wall, and extend the wall to the reconstructed appearance. The reconstruction work is intended to restore the section of the city wall to the appearance recorded in the historic photographs (when the wall was already substantially eroded); as a result the crenellations and parapet have only been partially reconstructed. The works also necessitated substantial work to the base of the wall to infill and repair the existing undercut, and install drainage away from the reconstructed wall base.

A very different instance of 'reconstruction' is the rebuilding of the Chapel of Reconciliation, Berlin (Germany) (*Appendix 6*). This work was concerned with the building of a new chapel on the site of the destroyed former church. The approach was to build a smaller chapel that reflected the declining congregation of the church. Important within the structure was the re-use of the building materials of the original church within the rammed earth walls. In this way the utilisation of the old building materials within the new structure allows the values and meanings associated with the historic fabric to be incorporated within the new building.

On other sites a combination of different approaches has resulted in the whole or partial reconstruction of sites and buildings, most commonly within the study area this was concerned with the utilisation of both encapsulation of the archaeological and historic fabric, alongside its partial reconstruction. At Rayen (Iran) the impressive citadel walls have been encapsulated using newly made mudbricks alongside the reconstruction of crenellations at the wall top, again using mudbricks, earthen mortars and earthen plasters (see above, *Appendix 6*). In a number of places this has involved the reconstruction of parts of the city wall, to raise the wall higher in order to create a complete circuit of crenellations. Where this has been carried out the newly constructed wall has been rebuilt using mudbricks, regardless of the original construction it replaces. Similar work combining approaches to encapsulation alongside the reconstruction of the upper layers of the city walls and crenellations was also recorded at Yazd (Iran), and in Uzbekistan at Khiva, Bukhara and Shahrisabz (see encapsulation above, *Appendix 6*). In all instances a combination of materials ranging from mudbrick, earthen mortars and plasters, through to breezeblock, and fired brick, covered in cement renders have been used.

Ex situ reconstruction sites

The experimental house at Çatalhöyük (Turkey) was built in the late 1990s-2000s as part of the ethnographic work on site (Stevanovic 1999) (Fig. 146-147, 156, *Appendix 6*). This reconstructed building is not based on a single excavated structure, but rather on an amalgam of different structures excavated on site. It uses a variety of different construction materials and techniques (some modern, (plastic sheeting) some perceived to be ancient (mudbrick, mudplaster, flat earth roof)). The purpose of this reconstruction is to assist in archaeological explanation, interpretation and understanding for both specialists and visitors to the site.

Restoration

Within the Arg-e-Bam (Iran) a number of structures have been restored using a combination of materials, such as fired brick, cement, lime mortar, mudbrick, earthen mortars and render (*Appendix 6*). The work in Arg-e Bam involved the construction of new roofs, and new wind catchers to recreate the look and feel of the infrastructure associated with the old town. As with the capping and encapsulation, similar materials and techniques are used through the site, with the use of both traditional and 'new' materials (Fig. 169-170). Generally fired brick and cement mortars were used for the core of the monuments, and then the finishing of this work with mudbricks, earthen mortars, renders and *gaunch*, to retain the look and feel of the original earthen architecture. A majority of these structures are restored with no adaptation or re-use planned, such as the mosque structure, central thoroughfare and gatehouses. Rather than having a function, these structures are designed to create the look and feel of the old town for visitors to the site. Other buildings had been restored and adapted to suit the needs of new uses, such as cafes, ticket office, bookshop, and ICHO office for Arg-e Bam (Fig. 188-189).

In Yazd (Iran) some of the restoration work is concerned with the stabilisation and encapsulation of the eroding historic earthen fabric within new mudbrick and *kahgel* (*Appendix 6;* Fig. 167, 168). This approach was used even in those areas in which the original construction utilised a *chineh* earth technique, such as within the Dowlat Abad Historical Complex. Similarly in Yazd a number of non-monumental structures within the old town have been restored for adaptive re-use (*Appendix 6*). The restoration work is normally associated with the adaptation of the structures for new uses. These uses are both 'local' (government and education buildings) and concerned with tourism (restaurants and hotels). For example the 'New Life for Old Structures' programme aimed to re-use historical buildings to meet the needs of a fast growing urban population in various locations in Iran. A number of these reused buildings were in Yazd. The Khan Bathhouse, has been reused as a restaurant; the Moayed A'layi House has been reused as offices (completed 1997), and Hosayniyeh Nazem ot-Tojar reused as an arts centre (completed 2000). In the programme buildings are acquired, restored and sold or let to new owners or tenants. The work was undertaken by the Urban Development and Revitalisation Corporation (UDRC), guided closely by the ICHO, this programme received an Aga Khan Award for Architecture in 2001 (Frampton *et al* 2001). The restoration approach has been to keep structural changes to a minimum, and to use traditional techniques and materials (replacement mudbricks, earthen mortars and plasters), alongside new materials (cements, drainage materials etc) where appropriate. The pragmatic approach is to ensure the restoration is cost effective, and by providing examples of work, ensure the methods used are transferable (*op cit*). Though the restoration work on these sorts of buildings has used a variety of different materials, such as concrete renders and mortars, and re-inforced concrete for structural repairs, the buildings are finished with earthen renders and *gaunch* (but more commonly with coloured cement-based renders), to create an appearance in-fitting with the urban core.

In Samarkand (Uzbekistan) a majority of the high status monuments have been restored in different periods (*Appendix 6*). The restoration work reflects the different pre-occupations of conservation and restoration through time. It is significant that for all of these different approaches the materials and techniques used for conservation were generally re-mortaring of brickwork using harder cement-based mortars, replacement of fallen bricks and tiles using new bricks and tiles manufactured to look old, substantial work at the base of the buildings to prevent problems associated with damp, and the capping of roofs with either cement or sheet metal. Emblematic of the restoration approach is the work carried out in the Registan Square (although primarily fired brick and glazed tile, earthen mortars and plasters were used in the original construction). Here the 3 *maddrassah*, present a façade of 'newness' and good repair to visitors to the site. However they are in a very poor state of repair, and a multitude of replacement materials have been used throughout (similar approaches were also recorded in Khiva, Bukhara, Shahrisabz, Anau and on some of the monuments at Merv, see *Appendix 6*).

Reconstruction and restoration assessment

Practical impact

Reconstruction and restoration can result in the long-term retention of sites and structures. Reconstruction can be of enormous importance in assisting with the interpretation and understanding of a site or buildings shape and form. Whilst restoration can work well in limiting erosion to earthen architecture (particularly when restoration restores damaged or fallen roofs), and the approach is particularly effective when new uses can be found for old structures (such as in Iran at Yazd and Arg-e Bam).

A variety of problems may be posed by reconstruction and restoration, particularly when inappropriate harder materials are used (see discussion in capping and encapsulation above). In addition *in situ* reconstruction utilising the 'original' fabric (often combined with other conservation approaches, such as encapsulation) can compromise the archaeological and historic fabric, particularly so as the work may damage the below ground buried archaeology, and other alterations in drainage and moisture can accelerate loss and deterioration.

Most problematic with reconstruction and restoration is the basis upon which the degrees, extent, form and shape of the intervention is decided. This may be based on historical evidence (such as photographs), archaeological evidence and interpretation, or fantasy. This is problematic as all of these interpretations are liable to change and fluctuate in the future. As reconstruction makes such a huge visual impact to a site, areas tend to become permanent and static features that do not reflect the current and changing interpretation. If in time reconstructed work is to be dismantled, it may be difficult to 'unpick' the work, and the damage that may occur to the historic of archaeological fabric can be substantial. For example, it is difficult to understand and interpret the reconstruction work carried out on the city wall of Abdullah Khan Kala at Merv. Here the historic photographs used for planning the reconstruction work recorded the already eroded condition of the monument. As the reconstruction work is intended to replicate the condition recorded in the historic photographs the height and form of the reconstruction work is not complete. However as this partial reconstruction is finished with earthen plaster the appearance of 'newness' alongside the expectation of reconstruction (as complete and whole) generates the perception that the wall is in a shape and form that would have existed during the past. As such of all the conservation and management solutions for earthen architecture reconstruction and restoration pose the greatest visual alteration to a sites shape and form.

There is concern over the type of information transmitted to the visitor through the *in situ* reconstruction of ruins (Sivan 1997) can be misleading and disorientating (Demas 1997; Schimdt 1997). This is particularly so when full *in situ* reconstruction is contrasted with other parts of a site in which there have been no interventions. As a result it can be difficult to understand why some areas of a site seem to survive 'untouched' in contrast to other poorly surviving areas of a site (in contrast to unrestored sites that fit seamlessly within the site setting). The same 'misleading' label may be applied to restoration. For example, in Samarkand (Uzbekistan) the restored monuments conceal their poor condition.

The restoration work within the Arg-e Bam (Iran) used a variety of different materials and techniques, ranging from traditional earthen mortars and renders through to the use of hard cement-based materials for mortaring and capping of the earthen materials. The problems associated with this work after the 2003 earthquake have already been discussed (in capping/encapsulation above). However, the work does show the importance of pragmatic responses to the retention and re-use of earthen architecture, utilising a variety of traditional and imported building materials and techniques. Despite this, problems exist in using the prescribed materials and techniques for restoration, in the Arg-e Bam all the restored structures tend to look exactly the same (and indeed the restored buildings in the Arg-e Bam looked the same as those in Iran at Yazd and Rayen). The result is to create an image of the Arg-e Bam as a single-phase site, ignoring the complexity of the sites shape and form, whilst using similar restoration approaches on other historic monuments in Iran substantially reduces local distinctiveness.

Current conservation theory

Reconstruction and restoration pose significant problems and controversy in relation to conservation theory. In order to avoid confusion reconstruction is kept separate from restoration in discussing conservation theory.

Up until the 1990s most international and national conservation charters raised concern over reconstruction. For example the 1931 Athens Charter advocates:

"In the case of ruins, scrupulous conservation is necessary, and steps should be taken to reinstate any original fragments that may be recovered (anastylosis), whenever this is possible; the new materials used for this purpose should in all cases be recognisable." (Article VI).

Whilst the 1964 Venice Charter advocates:

"All reconstruction work should however be ruled out "*a priori*." Only anastylosis, that is to say, the reassembling of existing but dismembered parts can be permitted." (Article 15).

However by the 1990s a more pragmatic approach is voiced in the 1999 Burra Charter:

"Reconstruction is appropriate only where a place is incomplete through damage or alteration, and only where there is sufficient evidence to reproduce an earlier state of the fabric. In rare cases, reconstruction may also be appropriate as part of a use or practise that retains the cultural significance of the place." (Article 20.1).

However this role can only be carried out if the work is: "Identifiable on close inspection or through additional interpretation." (Article 20.2).

Conservation charters also note the importance of *ex situ* reconstruction, and the contribution this makes to education and interpretation. The ICOMOS Charter for the Protection and Management of the Archaeological Heritage 1990:

"Reconstructions serve two important functions: experimental research and interpretation. They should, however, be carried out with great caution, so as to avoid disturbing any surviving archaeological evidence. Where possible and appropriate, reconstructions should not be built immediately on the archaeological remains, and should be identifiable as such." (Article 7).

This stresses the role that reconstruction can have in assisting archaeological understanding and interpretation (see also English Heritage 1999).

Despite the recommendations of conservation theory, *in situ* reconstruction is often justified as a means to fulfil the management and presentation needs of ruins, and many archaeological and historical sites around the world have been subject to some degree of reconstruction. Conservation theory would not advocate large scale *in situ* reconstruction utilising materials found on site, particularly when they have been quarried from sites elsewhere and transported for the reconstruction work. Such a practice was the norm on sites in the USA Southwest (where timbers and stone were transported between sites to enable reconstruction work) (*pers comm.* Dabney Ford). Such a policy places value on the reconstruction work rather than on the *in situ* conservation and retention of the 'original' fabric.

The most significant problem associated with *in situ* reconstruction is the dramatic impact on understanding and interpreting a site. At Nisa (Turkmenistan) the central reconstructed structure poses difficulties in interpretation and understanding the shape, form and survival of the site. It makes no sense that there is such a 'new' looking structure located within the eroded and eroding archaeological site, and significant conjecture has crept into the structure (such as with the roofs and ceilings). In addition the reconstructed *kiva* at Aztec Ruins (USA) has an enormous impact on the site. It is of note that the decision was taken not to plaster the exterior as the 'original' would have been, but rather leave it un-plastered in the way the ruins were found on excavation (Lister and Lister 1987). Both of these reconstructions give an inaccurate and very confusing interpretation, not as these structures would have been in the past, but a confusing blend of past and present. As such it can be seen that conservation theory sees most problems with these approaches when conjecture creeps in, for example the 1964 Venice Charter comments:

"It must stop at the point where conjecture begins, and in this case moreover any extra work which is indispensable must be distinct from the architectural composition and must bear a contemporary stamp." (Article 9).

The 1999 Burra Charter advocates quite a narrow definition of restoration as:

"returning the existing fabric of a place to a known earlier state by removing accretions or by reassembling existing components without the introduction of new material." (Article 1.7).

And comments:

"Restoration is appropriate only is there is sufficient evidence of an earlier state of the fabric." (Article 19).

Conservation theory is perhaps more flexible in discussing restoration, and does advocate the notion of re-using, and finding new functions for old structures as a means of conserving monuments (1964 Venice Charter, article 5). As such the adaptive re-use of a number of the structures in Iran in Yazd and Arg-e Bam fulfils the requirements of current conservation theory. Whilst the approaches to this work are pragmatic (mixing old and new), conservation theory would perhaps advocate more use of traditional materials and techniques and/or a more dramatic distinction between the two. The conservation approach taken within those restored structures has been to reconstruct the appearance in a single building phase.

In addition conservation theory places importance on all phases of a monument, and advocates restoration should respect different styles and periods of construction. For example the 1931 Athens Charter states:

"When, as the result of decay or destruction, restoration appears to be indispensable, it recommends that the historic and artistic work of the past should be respected, without excluding the style of any given period." (Article 1).

Similarly the 1964 Venice charter states:

"The valid contributions of all periods to the building of a monument must be respected, since unity of style is not the aim of restoration." (Article 11).

As can be seen from the sites already discussed significant tensions are posed when carrying out restoration work in order to keep the evidence from different periods and phases. As seen through the study area much restoration and reconstruction is carried out to a scheme without reference to enough information about a structures development or history.

Values of earthen architecture

Reconstruction and restoration can alter the perception of earthen architecture. For example within the Arg-e Bam (Iran) the decision to restore the historic fabric within newly manufactured mudbricks regardless of the earthen building technique utilised in the original, alters the interpretation of the structure. It suggests that in the past the only earthen building technique was mudbrick rather than placed earth, whilst the decision to replaster the surfaces of the earthen walls regardless of whether or not this had occurred on the original has significantly reduced the local distinctiveness associated with traditional uses of earthen architecture in Iran. The work also creates an 'image' and impression of the Arg-e Bam as a single-phase site, ignoring the complexity of the sites shape and form (indicating longevity and durability) and local distinctiveness of the use of earthen architecture.

If replacement materials are used, the visibility of these materials can generate and re-enforce the negative perception of earthen architecture as 'unconservable'. For those sites where harder, replacement materials are used, the values associated with a site are changed (effectively removing the earthen elements and replacing the soft contours of eroded earth with harder, more regular shapes). Similarly a number of the consolidants used in the reconstruction work on sites in the American Southwest change the colour and texture of earthen architecture. As such these approaches alter the aesthetic values associated with earthen architecture.

In other instances the utilisation of earthen elements, but lax maintenance policies can generate the perception that this is the condition in which earthen structures may have appeared in the past, again altering and re-enforcing the negative perception of earthen architecture. In the time since the reconstructed house at Çatalhöyük (Turkey) was built there has been no maintenance to the structure, and indeed what has been done - painting over the education project images with an synthetic household paint, is highly questionable. As such the understanding of the building to visitors who come to the site is altered, generating the perception that in the past these buildings were not maintained (conflicting with the archaeological evidence that indicates they were very well maintained (see *Chapter 6*). The impact on those still utilising earthen architecture in living contexts is perhaps more profound. If 'specialists' build a house that is then left un-maintained, this sends a powerful (and very high profile) message, absorbing from this that a lack of maintenance is acceptable, and that modern synthetic paints are an acceptable alternative to earth plasters. In the long-term the adoption of these attitudes reinforces the negative view of earthen architecture, in addition they will reduce the longevity of earthen structures, influencing further the rejection of earthen architecture and adoption of building materials perceived as more modern and more long lasting.

Other restoration work could utilise the qualities and properties of earthen architecture as a recyclable material, thereby allowing restoration work to be undertaken using the collapsed, and eroded earthen materials that can be recycled and reused rather than use new materials. Whilst this approach is problematic, and can falsify the archaeological record, it could serve both the practical role of ensuring new materials matched old, whilst exploiting and retaining the theoretical associations of the materials. For example, the Chapel of Reconciliation, Berlin (Germany) is interesting as it shows a radical departure from the reconstruction policies adopted for the reconstruction of other war-damaged monuments. The structure exploits the potential of earth to be recycled and to incorporate aggregate inclusions as a means to retain the connection between old and new buildings. This makes an interesting and thought provoking statement regarding the re-use and recycling of materials, alongside the retention of those meanings associated with the materials. The philosophy of the building and the choice of materials is a reflection of the perceived positive values of earthen architecture. The use of earth also makes reference to its perceived transient properties. In this instance the purpose has not been to build a structure that will 'last forever' but rather to build a structure that reflects the place and people who use it.

Sustainability

Reconstruction and restoration can reduce the sustainability of monuments and sites, as the materials and techniques used in the present, result in structures being passed on to future generations in a form and extent that past generations would not recognise. Of all the conservation approaches to earthen architecture reconstruction and restoration make the greatest impact on the archaeological and historic fabric in the present, and future generations will be left managing, understanding and interpreting (and if necessary re-interpreting, reversing and reworking) the legacy of conservation work.

The sustainability of the approaches is particularly problematic as much of the effort is invested in the initial outlay, without provision made for monitoring and maintenance in the long-term. As a result much of the reconstruction work is in a relatively poor condition. For example the experimental house at Çatalhöyük (Turkey) has not been regularly maintained and suffers from a lack of ownership, whilst the original construction of the experimental house reflected aspirations for the ethnographic work its subsequent use has altered the values, meanings and authenticity of the structure. In the long-term the lack of ownership (and hence maintenance) threatens its sustainability.

All restoration and reconstruction work increases the maintenance needs of a site or building, resulting in greater work, and greater need for funding to maintain the reconstructed building and/or element. This is problematic as on many sites the needs imposed by the buildings and or sites are already extensive without adding to the maintenance needs with reconstructed buildings. This tension is perhaps best illustrated by the reconstruction at Aztec ruin, where the reconstructed *kiva* is maintained, whilst the excavated archaeology has been reburied/backfilled as a means to reduce the financial burden of site management and conservation activities (see backfilling above). This means that only the 'reconstructed' elements of the site visible, whilst the 'real' elements are reburied.

Similar problems have been posed by different approaches to conservation, for example in Samarkand and the other sites visited in the study area, these approaches are unsustainable as they tend to be focussed on those monumental structures that are the most famous, ancient or iconic. The retention of the monument ensembles rests on the financial capacity for the restoration. The commercial nature of some of the restored monuments is directed towards a foreign tourism market (such as those in Samarkand (Uzbekistan). For these monuments the fluctuation in the tourism numbers determines the type of conservation work carried out, whilst at the same time high tourist numbers place the monuments at ever-greater risk.

In contrast the best examples of restoration and reconstruction can dramatically improve the sustainability of structures by both protecting them from environmental deterioration, and changing and installing new functions to structures that are valued, and retained for future generations (such as is the work in Yazd (Iran)).

Chapter 15: Removal/relocation

Removal/relocation is the taking away of a monument or site (or part thereof). Often removal/relocation is used alongside some form of consolidation to allow the display of objects or parts of a site within a museum. The purpose of removal/relocation is to remove material away from the most damaging effects of deterioration and destruction, and relocate it in order for them to be conserved, displayed and interpreted. The types of materials subject to this approach tend to be the most spectacular and high status discoveries made on a site, such as wall paintings (particularly when applied onto earthen plasters and earth walls which have been perceived as 'unconservable' *in situ*).

Removal/relocation has been a characteristic response to the problems posed by the conservation of excavated archaeological material, and on many early excavations the most high status and aesthetically pleasing material was immediately removed for conservation and display in museums of site (often around the world, rather than in the country of origin - infamously the Parthenon freeze at the British Museum, and the Gate of Ishtar at the Pergamon Museum). This approach is documented on numerous archaeological sites around the world, although increasingly the emphasis of archaeological site conservation and management places a preference on the conservation and display of materials *in situ* (albeit that this is also complex and complicated, and often results in impacts to a site through the erection of shelters).

Removal/relocation is an approach that is not particularly well documented in the literature concerned with the conservation and management of earthen architecture. Within the site dossiers removal/relocation is documented at Çatalhöyük (Turkmenistan), and within the study area at Samarkand (Uzbekistan) (Table 14). (*n.b* reference is only made to the removal/relocation of substantial parts of the site or building – such as whole paintings – not the other excavated artefacts, archives etc as seen at Nisa (Turkmenistan).

Sites

Removal/relocation is an approach characterised by the conservation work undertaken in the 1960s at Çatalhöyük (Turkey) (*Appendix 6*). The finds from the 1960s excavations were initially deposited with the Konya Museum, but later taken to Ankara, where they were conserved and displayed within a museum environment. A variety of techniques were used, including *strappo* (detachment of the paint layer alone), *stacco* (detachment of the painted surface including the underlying plaster surface), or *stacco a massello* (removal of entire walls). In addition polyvinyl acetate was used in the field on the very poorly preserved walls prior to removal. At Çatalhöyük the nature of archaeological excavation carried out in the 1960s, was such that the sequence within the open trenches was more-or-less completely excavated. Obviously this means that even the most dramatic of the wall paintings could not be displayed *in situ*. This excavation approach differs to that used today where the *in situ* conservation of the exposed walls has become a key aspect of the approach to site conservation and presentation, although some of the decorated wall plaster is removed in the course of excavation (Fig. 193).

The Sogdian wall paintings discovered on Afrasiab, Samarkand (Uzbekistan) in 1965 were removed and relocated, conserved and reconstructed in the excavated form and displayed in the Afrasiab museum (*Appendix 6*; Fig. 191-192). The paintings were discovered in the course of rescue excavation ahead of road construction, so in this instance there was no choice other than to remove, relocate, exhibit and display the paintings. The room in which the paintings are displayed is significant as it gives a sense of the scale, size, shape and form of the palace structure in which they were found, and allows visitors to follow the visual narrative through the room.

Removal/relocation assessment

Practical impact

Removal/relocation can allow materials to be conserved and presented without using other conservation interventions that could dramatically damage a site (or the materials).

Removal/relocation poses some problems primarily with the materials used for conservation and consolidation, as these may change the texture and/or colour of the conserved objects.

The most substantial problems posed by removal and relocation are associated with the interpretation and understanding of material that is no longer in context. This can make it difficult to understand the conserved materials that can appear as objects in an art gallery rather than integral parts of a site or structure. To some extent the decision to preserve and present the Sogdian wall paintings in Afrasiab (Uzbekistan) within a 'room' has had the effect of decontextualising the objects as 'art'. Problems are also associated with the materials used for consolidation and conservation of the objects. Today the condition of the Sogdian wall paintings within the museum is poor, and they have been damaged by the materials used in consolidation and the inappropriate materials used for installation.

Current conservation theory

The removal of material from sites and the relocation, conservation and *ex situ* display, poses problems associated with conservation theory. The removal and relocation of material from structures or brought to light in the course of archaeological excavation is only recommended as a 'last option'. For example the 1931 Athens Charter states:

"the removal of works of art from the surroundings for which they were designed is, *in principle*, to be discouraged." (Article V).

The 1964 Venice Charter reiterates this position stating:

"A monument is inseparable from the history to which it bears witness and from the setting in which it occurs. The moving of all or part of a monument cannot be allowed except where the safeguarding of that monument demands it … " (Article 7).

The same attitude of removal and relocation being a 'last option' is advocated through the 1999 Burra Charter (article 9.1)

The ICOMOS Charter for the Protection and Management of the Archaeological Heritage 1990 states more strongly:
"The overall objective of archaeological heritage management should be the preservation of monuments and sites in situ … Any transfer of elements of the heritage to new locations represents a violation of the principle of preserving the heritage in its original context." (Article 6).

Other conservation charters focus on the practical and ethical constraints in removing material (see the 1956 UNESCO Recommendation on International Principles Applicable to Archaeological Excavation, Article 8; Burra Charter Article 33).

It is interesting that conservation theory advocates the conservation of objects *in situ* in order to retain the context and limit further damage, as it is just this approach that has resulted in tensions between the current archaeological and conservation aims at Çatalhöyük (Turkey). Here *in situ* conservation is seen as limiting archaeological discovery and research, and therefore impacts the retention of the all of the values (particularly its future research potential) associated with the site (see site dossier).

Conservation theory similarly advocates only limited intervention on the archaeological or historic fabric. This is problematic as removal/relocation as seen at Çatalhöyük (Turkey) and Samarkand (Uzbekistan) often results in quite substantial intervention both in removing the material in the first instance and in subsequent conservation and display (for further discussion see consolidation above).

Values of earthen architecture
The removal and relocation of archaeological and historic material can generate and reinforce the negative view of earthen architecture, suggesting that there are no other solutions for conservation and management, generating and re-enforcing the notion of earthen architecture as 'unconservable'.

Many of the materials and techniques used for conservation and display change the colour and texture of earthen architecture, changing the material from the soft, matte-look to create a hard, shiny film on the surface. This presents a significant challenge to the positive aesthetic values of earthen architecture. Similarly these consolidation materials utilised for display alter the characteristics of earthen architecture from breathable (a positive attribute) to an 'unbreathable' material.

Sustainability
Removal/relocation can result in the long-term retention of the materials conserved and presented to the public in museum contexts, and this is particularly the case in contexts where materials have come to light during rescue excavations and/or the nature of archaeological excavation necessitates removal. This means that through removal and relocation materials are made available for future generations to visit, interpret and study.

However problems are associated with the longevity of some relocated materials, similar to the problems noted with the longevity of the variety of different materials used for consolidation (see above).

The removal/relocation of materials does contribute to a significant alteration to the site or building from which they come from. The alteration in the form and values associated with a building or a site, impacts both the present and future. This approach offers another conservation paradox: if all the material from a site were to be removed the associated values would change, meaning future generations may not value the site or building in the same way as it is valued in the present. For example if the high-status material is conserved and displayed in a museum, whilst the site has been subject to complete excavation they may be nothing to associate with the site, and so the focus and value of the site is lost in preference to the museum.

Chapter 16: Sheltering

Sheltering is a method of shielding a monument or site (or part thereof) against weather, normally through the erection of temporary, semi-permanent or permanent structures. Shelters are used on both archaeological sites, and over extant remains of historic earthen structures.

The purpose of sheltering is to protect the exposed site or structure from falling water, wind abrasion, and moderate annual and diurnal temperature fluctuations (and other local environmental erosion). Temporary shelters have been used during excavation on sites such as Çatalhöyük (Turkey) to moderate the environment in which exposure occurs, and to try to limit the rapid drying out which results in discolouration, cracking and exfoliation of the excavated earthen material almost immediately upon exposure. More permanent shelters moderate the environment in which exposed materials and sites are located in order to limit the damaging effects of climatic and environmental deterioration and erosion. On account of the foundations these types of shelter create considerable impacts on the below ground archaeology in an area that exceeds the zone in which the shelter will be erected.

Sheltering represents one of the earliest conservation approaches for earthen architecture used at Casa Grande Ruins (USA) at the turn of the 20th century (Matero 1999; Matero *et al* 2000). Since then an extraordinary variety of different approaches, materials and techniques have been adopted for sheltering of archaeological sites and monuments since the turn of the 20th century. During the later half of the 20th century sheltering has often been recommended for the conservation on archaeological sites (ICOMOS Charter for the Protection and Management of the Archaeological Heritage 1990). The documented research has been concerned with developing methodologies for sheltering, and is primarily concerned with issues such as materials, design, and monitoring of the shelter microclimate (Aslan 1997, CMAS Special Issue on Shelters on Archaeological Sites). For earthen architecture in particular, sheltering has often been recommended (Balderrama and Chiari 1995, recommendations of the Third International Symposium on Mudbrick (adobe) Preservation (1980)).

Sheltering work documented on earthen archaeological sites has occurred in Egypt, Israel, Italy, Peru, Turkey, USA and Uzbekistan. Within the site dossiers sheltering work is documented in Turkey at Çatalhöyük and Konya; sites in the Southwest USA (Aztec and Bandalier), and within the study area on monuments in Uzbekistan at Samarkand and Shahrisabz (Table 15).

The approaches to sheltering vary in terms of the perceived longevity and permanence, reflected in the various different materials and techniques used for shelter construction and design.

Low impact temporary shelters

These can be erected simply, and have only a limited impact on the belowground archaeology where the supports are placed. These shelters do not impact drainage, and can be easily assembled and dismantled at the completion of the excavation season. These shelters are designed to assist the archaeologists working under them, and provide some protection to the material by moderating the climate in which exposure occurs. For example the Polish trench shelter at Çatalhöyük (Turkey) consists of wooden support posts, upon which strips of tarpaulin are supported with guy-ropes (*Appendix 6;* Fig. 159). The shelter provides cover and shade for the excavation, and it is easily assembled and dis-assembled at the start and completion of the excavation by a team of local workmen. This shelter is good in providing shade, moderating temperature, and has a general low-impact. However without sides, the shelter provides very little protection from wind and sand blasting, so that the material exposed in the course of excavation can still rapidly dry out resulting in exfoliation and collapse.

Other types of temporary shelter include traditional low brush wood shelters, covered with vegetation matting that are erected on sites where conservation work involves the manufacture of new mudbricks. These shelters borrow from the types of structure erected in living contexts and allow more efficient mudbrick manufacture by moderating the drying out of newly manufactured mudbricks so that they do not crack as a result of drying out too quickly

Medium impact temporary shelters

These can be erected simply and have a supporting frame. These types of shelters have some impact on the ground, and are constructed from materials that tend to last a few years and/or offer protection from other environmental deterioration such as rainfall and wind.

At Çatalhöyük (Turkey) temporary shelters have the look of tarpaulin tent-like structures, but they have little impact on the below ground archaeology upon which supports are placed (*Appendix 6*). These shelters provide shade, cover and moderate temperature during excavation and, on account of the sides, also limit wind erosion and sand blasting. The more substantial build these shelters results in a greater impact on the visual and aesthetic aspects of the site. The BACH shelter at Çatalhöyük has a rigid metal frame weighed down by sandbags, and oil drums filled with concrete. The shelter over Building 5 at Çatalhöyük has a double skin that helps moderate UV and temperature on the inside of the shelter. The metal frame is weighed down using sandbags, and the overlap of tarpaulin at the base of the outer skin is an effective method of providing rainwater drainage (Fig. 160, 163).

At Aztec (USA) parts of the site have had various different types of *ad hoc* shelter or roofing structures (*Appendix 6*). The different methods used for the restoration of the roofs, include the use of cement, tarpaper, and felt. These reflect the development of different materials through the 20th century alongside the experimental use on archaeological sites. An early approach was to coat the excavated wooden roofs in poured cement, and in other places a wire barrier was installed and shellac coating applied (Lister and Lister 1987).

High impact 'permanent' shelters
'Permanent' shelters are of a more massive construction, and utilise more long-lasting materials in construction. These shelters create considerable impacts on the below ground archaeology in an area that exceeds the zone in which the shelter will be erected, and may not always be erected for the purpose of conservation (with some designed for sheltering visitors, Fig. 165).

At Çatalhöyük (Turkey) the south shelter is a concrete, steel and polycarbonate space frame structure covering the deep trenches on the southern slope (*Appendix 6;* Fig. 161-162). The shelter is designed to allow excavation to continue under shelter by moderating the climate in which excavation occurs, and also provide for the long-term display of the excavated material by stopping falling water and water run-off from eroding the exposed areas. Despite the archaeological limitations placed on the design, and methods taken to mitigate the impact of the shelter structure on the surrounding below ground archaeology this shelter required a 'rescue' excavation season to excavate and record the zone that would be impacted by the foundations and drainage channel. This shelter has had an enormous impact on the site setting and surroundings. In addition variations in drainage patterns and water-run-off in the zone surrounding the shelter, and variations in the climate inside the shelter, have caused further problems (see below for discussion).

The high impact 'permanent' shelter over the extant remains of the Konya Palace (Turkey) uses concrete to form an open parabolic curve that rises high into the air from the supports at the base (*Appendix 6*; Fig. 164). The shelter exploits developments in the design, manufacture, and realisation of the use of concrete in modern construction in the 1960s and 1970s. The shelter has an enormous impact on both the palace wall and its setting. By nature of its high profile location the modern design and materials have made a significant statement concerning the centre of Konya, a message concerned with the dual needs of conserving a historic past alongside adapting and re-developing a historic town. The shelter projects powerful messages concerned with this 'permanent' solution to the conservation of the palace, and by the nature of the materials and design utilised there is a clear message of separation between the past and the present.

Sheltering was also recorded at Bandalier National Park (USA); here one of the areas of decorated wall plasters from the *cavates* standing immediately adjacent to the visitor trial has been covered in a permanent long-term shelter (*Appendix 6*). This is in the form of a Perspex panel permanently affixed to the side of the canyon wall, allowing visitors to see the decorated plaster.

On the outside of the Gur Emir monument, Samarkand (Uzbekistan) the collection of decorated stonework has been protected by the erection of high-status permanent shelters (*Appendix 6*). The shelters comprise a metal frame covered in corrugated plastic/polycarbonate, and a barrier to dissuade walking on the surface. The same materials have been used to create a sheltered porch to protect the entrance to the monument. A similar shelter has been used at the Ak-Serai palace, Shahrisabz (Uzbekistan) to protect a decorated marble floor surface (*Appendix 6;* Fig. 166).

Sheltering assessment

Practical impact
The use of shelters is in some respects one of the most effective policies for excavated earthen architecture as it protects exposed structures from falling water in the form of rain or snow. This means the excavated or extant earthen architecture can be protected and displayed, and retains the visibility of construction, phasing and form. However the erection of a shelter can often change other aspects of the site (such as drainage and unexcavated archaeology), generate a microclimate, impact the visual component and aesthetics of a whole site, and the zone of excavation (by changing light levels, temperature and humidity).

The erection of shelters results in an enormous visual impact to an archaeological site or structure. At Çatalhöyük (Turkey) the shelters have altered the shape and form of the tell. This has been to the detriment of the whole site, as this encourages the notion that the shelters are erected over the most important, high-status zones, perceived as the 'only' significant parts of the site. As the shelters used on site all vary in design and form they also create problems by giving a sense of separate excavated zones rather than of a 'whole' site. The visual impact of these shelters at Çatalhöyük is extreme, impacting the natural shape, form and visibility of the tell. But perhaps this view of the site, as preserved and presented under shelter is suitable for an archaeological excavation in process, as it presents the idea of an artificial process that makes a huge impact on the tell. Excavation itself goes down, revealing the buried structures, whilst the shelters go up, protecting those buried deposits, and making the act of excavation visible to everybody.

The materials used for shelter construction are all subject to erosion and deterioration. At Çatalhöyük (Turkey) problems are apparent with the longevity of the plasticized shelter material used on the semi-permanent shelters, particularly where water has infiltrated through the material resulting in the growth of moulds, and damage has occurred as a result of a lack of maintenance (see below for further discussion). The shelters have created and posed drainage problems and increased the risk of loss and damage to the surrounding unexcavated archaeological strata through surface run-off, erosion and gullying (made worse as people walking around the tell, the excavated areas and the sheltered buildings cause further erosion to the loose friable surface).

Similarly the poor condition of the materials used for the Konya (Turkey) shelter attests to the limited lifespan of concrete and limited maintenance of the structure through time. The concrete is suffering particularly badly from shearing and the detachment of the outer and inner surfaces. The inside of the dome has also suffered from corrosion, possibly caused by water infiltration, pollution from passing traffic and/or corrosion of reinforcing metal elements used in the original construction. The birds that nest in the void between the different outer and inner concrete domes now contribute further damage. It is hard to see how this structure can be repaired (and/or dismantled) without causing undue damage to the surviving parts of the palace.

The degree of permanence to which a shelter is designed and constructed has an impact on the type, and extent of foundations required for the structure. The installation of foundations can have an enormous impact on the unexcavated below ground archaeology, and so by trying to protect one aspect of a site, a value is placed on this above the archaeology that is to be removed in the course of 'rescue' excavation. As seen at Çatalhöyük (Turkey) the type of foundations required can impact not just the area immediately under threat from the development, but also the surrounding archaeological deposits through the extensive alteration of surface drainage patterns. This places a value on the material to be protected within the shelter rather than on the surrounding, in most instances, unexcavated archaeological deposits.

The erection of a shelter may also damage the archaeological and historic fabric being protected. The *ad hoc* shelter measures used at Aztec (USA) have caused considerable damage to the timber roofs. For example, following the excavation of Room 117 in 1920, a roof was constructed comprising a wire barrier and shellac coating, however the shelter created a heating and evaporation differential (Lister and Lister 1987).

Shelters may also generate a microenvironment. As seen at Çatalhöyük (Turkey) shelters have a tendency to change and alter the microenvironment, impacting factors such as wind erosion and sand blasting, this affects surface run-off, and UV damage to the exposed earthen architecture and shelter construction materials.

Current conservation theory

As with backfilling, sheltering of archaeological sites and monuments has been recommended since the earliest discourses on appropriate solutions for conservation and management on archaeological sites (see *Chapter 7*). Contemporary approaches to site management and conservation nearly always make some reference to sheltering as an effective and appropriate preventative conservation tool, and considerable research has been directed towards shelter designs and materials (see CMAS Special Issue on Shelters on Archaeological Sites).

Unlike many of the other interventions shelters are identifiable as new work (Burra Charter Article 22.2). This makes these types of interventions particularly suitable for use on sites and structures, fulfilling the requirement of minimum impact on the archaeological or historic fabric, whilst retaining visibility and authenticity.

Unfortunately nearly all shelters have a wider impact on a site and can damage the unexcavated archaeology, impact a sites interpretation and understanding, and change the sites aesthetics. This is significant as current conservation theory would advocate the retention of all of the sites values and significance (Burra Charter Article 1).

The most significant impact of a shelter is on a sites setting. Article 8 of the Burra Charter states:

"*Conservation* requires the retention of an appropriate visual *setting* and other relationships that contribute to the *cultural significance* of the *place*.
New construction, demolition, intrusions or other changes which would adversely affect the setting or relationships are not appropriate."

As such the methodology at Çatalhöyük (Turkey) adopted for the South Shelter is problematic. This shelter is massive and when contrasted with the less permanent shelter methods used on site, the visual impact and lack of similarity in design, materials and construction makes the shelter over dominate the site. Similarly, the Konya (Turkey) shelter is a very visible intervention intended to protect the surviving parts of the monument. The shelter has enabled a minimum of intervention on the historic fabric, and so the different materials and construction details are visible without any modern intervention on the fabric itself. However it's removable may pose significant problems.

Shelters can significantly change the value and significance of a site or building. At Bandalier (USA) the approaches to displaying the wall paintings behind the Perspex panel make an enormous impact on the side of the cliff. It also suggests this is a work of 'art' to be viewed in a gallery rather than a functioning part of the cliff dwellings. Similarly at Çatalhöyük (Turkey) there is a tendency to see the site as a series of interventions protected by shelters, rather than as a complex 'whole' site. This danger makes the use of multiple different shelters on an archaeological site particularly problematic as it limits the understanding of a sites cultural significance gained through appreciating setting and context. Current approaches to conservation would perhaps seek a less intrusive design and/or greater similarity in the interventions.

Values of earthen architecture

The installation of shelters attests to the understanding of the properties of earth as a building material, in which erosion occurs in relation to environmental factors and fluctuations. As this is an 'environmental' solution it also reflects the positive values of earthen architecture as an environmentally friendly material.

By protecting earthen architecture from the damage caused by falling water, shelters can assist in retention and re-enforce the positive perception of earthen architecture as a durable material. Similarly by offering protection to earthen architecture without recourse to other materials and techniques, sheltering can retain the visibility of the construction coursing, and phasing of earthen architecture, reinforcing positive values of durability, ancientness, and diversity.

The use of concrete as a material for the Konya (Turkey) shelter represents complex perceptions of the qualities of concretes in contrast to earthen materials. As we have seen concrete and cement products have often been perceived as the 'permanent' solution to the problems posed by the retention, conservation and management of earthen architecture. The failure of the concrete used in the Konya shelter is unsurprising, but the shelter and the palace it preserves stands as an amazing set of oppositions: the new shelter using 'modern' materials, eroding and deteriorating so quickly; as opposed to the historic fabric which has retained much of its form through several centuries of abandonment and earthquake damage.

The use of shelters for the conservation and management of earthen architecture results in a substantial alteration to the shape and form of a site or building. The positive adaptable and aesthetic values associated with earthen architecture, are contrasted by the hard, rigid shelter structures.

Sustainability
The use of shelters for the conservation and management of earthen architecture can result in the retention of materials, and passing them on to future generations in the current form. As such sheltering is one of the more sustainable approaches for the conservation of earthen architecture.

As with the other approaches used for the conservation and management of earthen architecture, the sustainability of sheltering can only be assured if monitoring and maintenance occurs. These requirements raise issues concerned with the economic sustainability of this approach, where too often funding and political will is available for the initial construction of the shelter, but the longer term maintenance (and in time replacement) is much more complex.

The sustainability of this approach is also questioned by the design and types of materials used for sheltering. For example the harder cement-based materials used for the Konya (Turkey) shelter threaten the sustainability of the palace. This shelter shows how problems have been posed by the lack of longevity of 'modern' building materials used for conservation. As a result this approach can raise the possibility of handing on to future generations the legacy (and conservation demands) of the 'shelter' rather than the site or structure the shelter was originally intended to protect.

Summary

The above discussion and data collected from the site dossiers is intended to cover the majority of the different approaches for the conservation and management of earthen architecture in a variety of different contexts. However, a number of other approaches are noted within the documented research concerning the conservation and management of earthen architecture, but to a lesser degree within the site dossiers.

Perhaps the most common approach not discussed is stabilisation. Stabilisation is a method of making a monument or site (or part thereof) more durable - such as a substance that retards chemical action or improves resistance to altered loads. Of the approaches utilised for earthen architecture stabilisation is most often combined with other approaches, such as reconstruction and restoration. Stabilisation is for both structural and seismic purposes. Although many of the monuments in the study area have been subject to some form of stabilisation work, this is often, as is the case with the monuments in the Registan ensemble in Samarkand (Uzbekistan), as an integral part of other approaches, such as restoration and reconstruction.

Chapter 17 returns to my research goals summarising my understanding of current approaches to the conservation of earthen architecture; establishing a transferable intellectual framework for earthen architecture on archaeological sites; and understanding 'difference' in approaches seen within conservation and heritage theory. The first part of *Chapter 17* will summarise and critique the conservation approaches recorded from the different sites synthesised within this chapter. This summarises the effect of the different conservation approaches, materials and techniques in light of practical impact, current conservation theory, values of earthen architecture and sustainability. The chapter concludes by answering my research questions and highlighting areas of potential for future research concerned with earthen architecture, conservation and heritage theory.

Chapter 17: Critique and Discussion

This chapter returns to the research goals and questions established in *Chapter 1*. The first part of this chapter addresses my first research goal - to develop an understanding of current approaches to the conservation of earthen architecture. This summarises and critiques the conservation approaches recorded from the different sites synthesised within *Chapters 8-16*, looking at the effect of different conservation approaches, materials and techniques in the light of practical impact, current conservation theory, values of earthen architecture and sustainability.

The second part of this chapter addresses my second research goal - to establish a transferable intellectual framework to assist in the conservation decision-making process for earthen architecture on archaeological sites. This draws upon all aspects associated with the conservation and management of earthen architecture highlighted in this work.

The third part of this chapter addresses my third research goal - to develop an understanding of 'difference' in approaches seen within conservation and heritage theory. This uses the data collected through this research to understand contemporary issues within conservation and heritage theory concerned with the identification of 'difference' and 'otherness' in relation to the conservation approaches for a single class of material: earthen architecture.

In conclusion, I highlight areas of potential future research in the conservation and management of earthen architecture, and reflect more personally on the research.

Conservation approaches for earthen architecture

This research has been concerned with developing an understanding of current approaches to the conservation of earthen architecture, focusing upon the impact of the different materials, techniques and approaches, the relationship to current conservation theory, and the values of earthen architecture and sustainability (see summary Table 4).

Practical impacts

In the first instance, all of the different approaches to the conservation and management of earthen architecture *can* assist in prolonging the life of structures and sites. On many sites a number of different conservation approaches have been combined, such as backfilling and drainage, consolidation and sheltering, and undercut repairs and drainage. These combined methods often work particularly well because each conservation approach is effective in counteracting the erosion and deterioration caused by a single factor (such as shelters protecting from falling water), but additional methods are required for protecting the site or structure from other forms of erosion and deterioration (such as drainage work or undercutting repair).

Nevertheless, the different approaches, materials and techniques will seldom bring a complete halt to all of the factors associated with the deterioration of earthen architecture. As this research has identified, the erosion processes affecting earthen architecture are continuous (*Chapter 5)*, and methods used to retain earthen architecture must be monitored and maintained to be effective. Similarly all of the conservation materials, techniques and approaches have a significant visual impact that can influence the understanding and interpretation of a site or structure (*Chapters 8-16*). In other instances the conservation approach can have both positive and negative impacts across the whole site or structure, sometimes assisting in the retention of a particular zone or set of values but to the detriment of others.

As seen in *Chapters 8-16* specific conservation approaches create specific practical impacts. For example, there are problems associated with tying-in encapsulation work, whilst chemical consolidation generally works well *ex situ* or *in situ* when protected by shelters, whereas the use of cement-based renders creates a hard impermeable barrier below which there is an increased rate of erosion of the earth wall. On those sites where the conservation and management solutions have not been successful it is often associated with single interventions designed to protect against a single source of loss and deterioration, without necessarily planning for protective measures against other forms of erosion and deterioration. In other instances the installation of one form of protection can actually encourage and accelerate erosion caused by other factors (for example drainage problems, wind erosion and the generation of a microclimate may all result from the installation of a shelter - as at Çatalhöyük (Turkey) (*Chapter 16*)). This underlies the fact that conservation and management activities must be approached holistically, looking at *all* of the site, and *all* of the erosion and deterioration factors.

Often quite inconsistent approaches are adopted for the conservation and management of earthen architecture over the whole site. For example the use of both backfilling and *in situ* reconstruction on archaeological sites (as at Aztec (USA) - see site dossier and *Chapter 8*), these approaches present an uneasy tension, between on the one hand burying and making invisible the 'real' historical and archaeological fabric, whilst creating the 'new' reconstructed fabric. This runs the risk of subverting the values of an archaeological site or historic building, through inventing and re-inventing real and reconstructed histories. On sites such as Aztec (USA) the multitude of approaches is a reflection of the context within which the original interventions were undertaken, and highlight the problems of long-term conservation and management where each generation is left with the legacy of interventions carried out by the proceeding generation. Today problems such as these should be considered within the framework of sustainability, considering the impact of the past, present and future of the conservation approaches (see below).

Conservation Approach	Practical impacts	Current conservation theory	Values of earthen architecture	Sustainability
Backfilling	Protects site Visual Impact Can cause damage	Advocated through current conservation theory Minimum impact	Changes appearance and aesthetic values Can seem 'unconservable'	Needs monitoring and maintenance
Capping and Encapsulation	Protects site Visual Impact Can cause damage	As a maintenance activity advocated through current conservation theory, minimum impact. Encapsulation is not advocated by current conservation theory and can have a big impact	Changes appearance and aesthetic values. Use of harder material can seem 'unconservable' Reduces local distinctiveness	Needs monitoring and maintenance
Consolidation	Protects part of site Can have visual impact Can cause damage	Advocated through current conservation theory Minimum impact	Changes aesthetic values Can seem 'unconservable'	Needs monitoring and maintenance
Do nothing	Does not protect site No visual impact	Not mentioned by current conservation theory Minimum impact	Changes aesthetic values Can seem 'unconservable'	Needs monitoring and maintenance
Drainage and Undercut repair	Protects site Visual impact Can cause damage	As a maintenance activity advocated through current conservation theory Minimum impact	Changes aesthetic values Can seem 'unconservable'	Needs monitoring and maintenance
Maintenance	Protects site Visual impact	Advocated through current conservation theory, for 'living' sites Minimum impact Alters 'age value'	Changes aesthetic values Appropriate to renewal values	Needs monitoring and maintenance
Reconstrucion and Restoration	Protects site Visual impact	Not advocated by current conservation theory - unless new role for structure Big impact	Changes aesthetic values Reduces local distinctiveness use of harder material can seem 'unconservable'	Needs monitoring and maintenance
Removal /Relocation	Does not protect site Decontextualised	Only advocated by current conservation theory if 'last option' Big impact	Changes aesthetic values	Needs monitoring and maintenance
Sheltering	Protects site Visual impact Can cause damage	Advocated through current conservation theory Minimum impact	Changes aesthetic values Can seem 'unconservable'	Needs monitoring and maintenance

Table 4. Summary of impacts of conservation approaches.

Conservation approaches to earthen architecture such as the provision of drainage, reconstruction, sheltering, and undercut repair can have a considerable archaeological impact, through the removal of archaeological deposits, and they can also impact buried archaeological deposits through the alteration of groundwater (surface drains, for example, may work so well as to increase aridity). As a consequence archaeological deposits (and the associated material culture) can be damaged. Such is the case that understanding the importance of the archaeology of conservation and the conservation impact of archaeology should be seen as important factors in conservation and management planning for a site (see below).

Similarly, the survival and deterioration of sites is affected by the nature and context of archaeological work carried out. In some instances the location of archaeological excavations encourages the creation of conservation problems. For example, in the case of the archaeological excavation of tell sites, trenches have often been located cutting into the side of the tell (to reach older deposits), but such trenches disrupt natural drainage patterns and can encourage the more rapid erosion of the excavated material through the erosive effects of surface run-off. In other instances commencing archaeological work can impact already conserved areas through the alteration of drainage patterns. The practical impacts of archaeological work necessitate the requirement for holistic site conservation and management planning, providing linkage between archaeology and conservation and *vice versa*.

Most approaches to the conservation and management of earthen architecture result in significant visual changes to a monument, site or setting, and these can impact site understanding and interpretation. For example, backfilling means trenches are no longer visible and the excavated sections no longer legible; whilst sheltering means structures are added to the site, and encapsulation and reconstruction alter the shape and form of a structure. The visual impact of conservation activities can be moderated and managed through the utilisation of interpretation programmes to explain the nature and approach to the conservation of earthen architecture on a particular site.

In summary, all of the conservation approaches have practical impacts on the site or structure. They can assist in assuring the longevity of excavated or extant earthen architecture, in other instances ill-planned conservation approaches may contribute to more rapid erosion and deterioration. Most of the conservation approaches are responsible for wider changes affecting the site setting and visual characteristics of a site or monument. Some of the approaches, such as backfilling, maintenance and sheltering, often produce positive results in the long run, generally within the ethos and character of current conservation theory, but alter the visual characteristics of the site or structure. Other approaches, such as encapsulation, restoration and reconstruction, can also produce positive results in the long run, but significantly alter a site and challenge conservation theory. Approaches such as consolidation have generally produced negative results, but fulfil the requirements of current conservation theory. This perhaps highlights the conflicts apparent in site conservation and management, making explicit that this is a discipline concerned with balancing different practical needs and theoretical demands.

Current conservation theory

Chapter 7 defined a number of concerns of conservation theory, such as the importance placed on the archaeological or historic fabric, age value, visibility and reversibility, anti-restoration, authenticity, new materials and a role for science and industry, international importance, alongside those more current concerns of cultural and intangible heritage, such as values, participation and poverty reduction. *Chapters 8-16* have explored the debates associated with current conservation theory and the different approaches to the conservation and management of earthen architecture. In most instances the 1999 Australia ICOMOS Burra Charter provided a framework for assessing approaches in light of contemporary conservation theory.

Broadly, conservation theory influences what is valued and subsequently conserved, alongside the assessment of the appropriateness of the materials, techniques and approaches utilised. The 1999 Australia ICOMOS Burra Charter defines the process of conservation as concerned with the retention and management of all of a sites cultural significance (Burra Charter Article 1). However all of the conservation approaches have an impact on a sites significance, sometimes resulting in the retention of some of the sites values over others (such as with the case of sheltering which impacts unexcavated archaeological deposits, or restoration which often seeks to return a structure to its appearance in a single period). This illustrates how considerable and irresolvable problems and paradoxes exist when carrying out conservation work and meeting the demands of conservation theory.

Trying to meet the demands of conservation theory has resulted in the enormous diversity of materials and techniques utilised for the conservation and management of earthen architecture. A number of the approaches such as backfilling, consolidation and sheltering have developed in response to the notions such as 'minimum intervention'; others fall into a broad category of maintenance activities, such as capping, drainage and undercut work; whilst approaches such as encapsulation, reconstruction, restoration and removal are not advocated by current conservation theory.

Practical problems are posed by meeting the ideals created by conservation theory in relation to the physical properties and values of earthen architecture. For example reversibility (cited in the Burra Charter Article 15.2) of some conservation work can be problematic: earthen materials can blur the distinction between archaeological/historic fabric and conservation work, whilst consolidants and cement based materials can be difficult to remove. In other instances visibility (cited in the Burra Charter Article 22.2) of new work can be difficult to achieve as new work (particularly if it is to be plastered) is likely to be indistinguishable from old, the use of the same materials means it may be impossible to undo what has occurred, and erosion will blur the distinction between new and old work.

Trying to meet theoretical demands may impact upon the practical effectiveness of interventions. For example, in Bam (Iran) conservation work was carried out without the use of straw as a binder in newly manufactured mudbricks in order for the work to be visible and reversible. However the omission of straw reduced the effectiveness of the conservation work and may have been a factor that contributed to the destruction associated with the Bam earthquake (see site dossier). Similarly work at Merv concerned with the separation between the archaeological/historic fabric and the conservation material resulted in some experimentation with different geotextiles, designed to act as a separator between the eroded wall fabric and new earthen plaster layers, but these experiments had only limited success (see *Chapter 9*). In all of these instances attempts to provide for reversibility through separation have resulted in substantial negative impacts on the longevity of the conservation work. In other instances it is impossible to provide any sort of separator layer without sacrificing the integrity of the conservation work, such as with earthen materials used for capping, and earthen materials used for plugging drainage gullies. In these instances the success of the intervention rests on the need for cohesion between the new conservation materials and the historic or archaeological fabric.

Particular concerns of conservation theory have a significant impact on approaches to earthen architecture. Going back to the origins of the conservation debate, we can see that value was placed on the surface appearance of the historic fabric as a visible and truthful testament to the buildings life:

"in the course of this double process of destruction and addition the whole surface of the building is necessarily tampered with; so that the appearance of antiquity is taken away from such old parts of the fabric as are left" (Manifesto for the SPAB, 1877).

This emphasis is still present within the Burra Charter:

"Preservation protects fabric without obscuring the evidence of its construction and use."
(Burra Charter Article 17)

The importance placed on the surface appearance and age value of a site or structure raises particularly complex issues in relation to earthen architecture. For earthen buildings approached in a living context the earthen surface will annually be re-plastered. This re-plastering of the surface finish gives the aesthetic and visual impression of newness regardless of the age of the historic fabric. We have seen that this approach is viewed as one that "obliterates their conservation." (Çatalhöyük Management Plan 23). In this instance the theoretical divide between what conservation theory advocates in relation to the visibility and age value is in contrast with the physical properties and values of earthen architecture (also see Burman 1999; Warren 1993, 1999 for comment).

One result of this 'divide' between conservation theory and practice is the enormous amount of experimental work concerned with consolidants. The notion of the importance of the visibility of the archaeological and historic fabric has led to efforts to discover a consolidant to preserve the appearance, visibility and character of the original earthen substrate and phases of construction. Despite this being a focus of much of the conservation science research into earthen architecture from the mid 20th century onwards, it is the case that these consolidants have achieved little real success in use on site (see *Chapter 10*). Attempts to conserve without an earthen surface finish are problematic; these fulfil the theoretical demands of preserving the archaeological or historic fabric and age value of the monument through minimum intervention and serve a didactic function in preserving the visibility and narrative function of the archaeological or historic fabric. If successful, these attempts would be useful on archaeological features, and particularly on archaeological sections (as shown in Building 5 at Çatalhöyük); however, when attempted on buildings this approach ignores established patterns of maintenance and repair associated with the building function and integrity, as they aim to conserve earthen architecture in a form in which it would never have appeared. Similar debate has arisen in terms of the use of renders and re-pointing on masonry structures where both actions alter the surface appearance.

Maintenance using earthen plasters and renders can be considered appropriate if the structures are still occupied, and if the materials and techniques used are appropriate (Venice Charter Article 4; Burra Charter Article 16). In other instances maintenance may be an appropriate conservation solution for archaeological sites and for abandoned historic buildings (but only for those that were originally coated in an earthen plaster). These types of interventions can be quite confusing for the visitor so in these instances the intervention must be interpreted in order to avoid confusion and misunderstanding; understanding the appearance of 'newness' regardless of the age of the structure. Problems are posed when the materials and techniques used for maintenance alter (through a loss of knowledge or through the introduction of new materials); when this happens these altered patterns of maintenance may loose authenticity and be inappropriate for the conservation of earthen architecture, even in living contexts.

The 1964 Venice Charter and 1972 World Heritage Convention introduced the notion of authenticity, and this debate was further extended and explored within the 1994 Nara Document (*Chapter 7*). As such we can understand authenticity in relation to earthen architecture through craftsmanship and workmanship, materials, techniques and setting. Even with the post-Nara understanding of authenticity problems are associated with the concept, particularly as some traditional 'authentic' practices may actually encourage erosion of earthen architecture. For example Damluji (1992, 138) notes a tradition of placing salt in the foundation course of mudbrick buildings in Yemen; the traditional explanation is that the salt deters rising damp and insect activity, although the placement of salt may actually accelerate erosion. To revise this traditional practice in light of scientific knowledge of damage caused by salt to earthen materials questions the retention of traditions, and the authenticity of craftsmanship and practices associated with earthen architecture.

Conservation approaches increasingly understand the importance of maintenance and specifically the retention of traditional systems of maintenance and renewal (such as the 2004 INTACH Charter in India). Maintenance is reliant on social, economic and political contexts that retain use values associated with the archaeological and historical environment. When buildings fall out of use, normally as a result of social, economic and political change, maintenance regimes stop. If accompanied by other changes, such as the introduction of modern building materials, the skills associated with earthen architecture can be lost in just one generation. Several heritage organisations have emphasised the importance of training programmes in the maintenance and conservation of earthen architecture in different locations around the world, but often 'outsiders' drive these initiatives, and they present interesting challenges to the notion of authenticity and may threaten to 'invent' tradition (see Hobsbawn 1983; 2004 INTACH Charter in India).

Article 15 of the 1964 Venice Charter introduced the concept of *anastylosis*: the reassembling of existing but dismembered parts (*Chapter 7*). *Anastylosis* is challenging in relation to earthen architecture as original fragments fallen from an earthen building will, in time, weather and erode, leaving a deposit of earthen material which will require reworking prior to reinstatement. The reworking of fallen and eroding material is a preferred option for assuring similar properties between the conserved material and the conservation material (with the recording of cultural material present as inclusions, and with the addition of organics that may have been present as inclusions in the original construction but which have subsequently decayed). However, with such repairs for earthen architecture the area conserved will merge into the rest of the structure, leading to invisible repairs. At its most extreme the re-use of earthen building materials, such as within the Chapel of Reconciliation, allows a new building to be constructed on the same site, or away from the site of the structure, enabling the values associated with a site to be retained and re-interpreted within the materials utilised for a new structure (see *Chapter 14* & site dossier).

The conservation and management of earthen architecture has been described as "a very difficult one" (Abdurazakov 1986), which coupled with the generally negative attitudes to earthen architecture (see Fielden 1994, 73; Chapter 4), has led to the notion of the materials as 'unconservable'. This notion of 'unconservability' continues to be re-enforced by approaches to the conservation of the material that do not take into consideration the physical properties, values and associations of earthen architecture (see below). In conclusion, I would argue that earthen architecture is not 'unconservable', only that the material does pose problems related to the interpretation of current conservation theory. I would argue that we would be better placed as a discipline if we questioned what we mean by the conservation of earthen architecture which, for me, is concerned with the conservation of the values of earthen architecture (see the framework below).

At Merv, for example, we have been concerned with compromise between practical actions to conserve sites, weighed against the demands of conservation theory. There is a problem with transmitting the notion of compromise, especially for those sites and countries that have been criticised in the past for the nature of the conservation work (such as Uzbekistan) where recommendations by international bodies still highlight the lack of knowledge concerning 'international' principles and recommendations (*pers comm.* David Gandraeu – Central Asian Earth Initiative Recommendations). This critical approach is problematic as (1) there are still no real principles and recommendations specifically designed with earthen architecture in mind; (2) current international principles and recommendations are more concerned with processes, such as management planning, rather than actual conservation practice; and (3) where they do exist international principles and recommendations are interpreted as absolute and un-compromising regulations. For example, recent discussions at Merv highlighted considerable confusion amongst park staff concerning the practical application of conservation theory. For the conservation of the palace complex in Shahriyar Ark the efficacy of utilising a fired brick buttress to support a mudbrick wall was discussed (Fig. 68) and park staff envisaged this as appropriate, within the context of conservation theory, as the new material could be clearly distinguished from the old. This is significant, for while the notion of visibility is implicit within conservation theory, the application needs to be decided on a site-by-site basis to assess the appropriateness. In this instance the intervention fulfils the didactic notions of visibility, but performs poorly, disrupts the existing plan of the structure, and presents an aesthetic challenge for visitors to the site.

I would argue, therefore, that to usefully assist in planning for the conservation and management of earthen architecture we need a substantial re-think. This research, alongside the transferable intellectual framework for earthen architecture developed below goes someway towards that, by reviewing and critiquing what has been done in the past to arm practitioners with the knowledge of 'what we do' but within the context of understanding the practical impacts, interpretation(s) of conservation theory, values of earthen architecture and notions of sustainability (see below). Conservation theory should be included as one of the decision-making tools for earthen architecture as it provides the context for successful practical applications, but conservation theory is just one of many aspects that must be considered when planning for the conservation of earthen architecture.

Values of earthen architecture

Approaches to the conservation and management of earthen architecture both impact upon, and are impacted by the negative and positive values associated with the material explored through this research (*Chapter 4*). The negative view of earthen architecture is one that sees the material as lacking modernity, associated with poverty, backward and uncivilised, cheap, weak, more liable to destruction, linked to ill health and disease, a last resort, and one with unsuitable language associations. The negative perception of earthen architecture has significantly impacted the archaeological and conservation approaches to the material, further re-enforcing the perception of the material as weak and 'unconservable'.

More recently the positive values of earthen architecture have been explored (*Chapter 4*). The positive view of the material is one that is adaptable, aesthetically rich, ancient, autonomous, healthy, locally distinctive, linked to humanity, modern, resistant to environmental disaster, environmental friendly and responsive, and associated with a rich symbolism. Successful conservation can promote these positive values and associations of the material, particularly where a solution is utilised that enables the resource to be retained and interpreted. For earthen architecture such 'positive' approaches to conservation can be the utilisation of maintenance and shelters. However, many of the conservation approaches to the material do not reflect this positive view of the material, and in a number of instances actually negate or destroy the positive associations.

The physical properties and values associated with earthen architecture do make retention in its 'as found' form difficult (see *Chapter 4 & 5*). An extant or excavated earth structure will erode more quickly than a stone structure. This may mean that, in some contexts, approaches that at first may seem to compromise conservation theory can actually assist in retention and challenge the negative values associated with earthen architecture, for example some of the encapsulation/restoration work recorded in the study area do emphasise the durability of earthen architecture.

Conservation can reinforce the negative view of the material when the work has only a limited success. Sometimes work has suffered from a poor understanding of the physical properties and qualities of the material. For example, the decision to manufacture new mudbricks for encapsulation and restoration work without the addition of straw, in order to fulfil the conservation requirements of visibility and reversibility, limits the durability of the conservation work. The poor survival of the conservation work further reinforces the negative perceptions of earthen architecture (for example Bam (Iran), *Chapter 14*).

Abandoned earthen material is subject to erosion, deterioration and subsequent formation and deformation. This may make material available for quarrying and re-use in new construction (and conservation) work. In this context, conservation approaches that seek to retain a material with positive, recyclable attributes in an 'as found' form can be queried (and I would argue this is a much more complex issue for earthen materials that loose form over time, when compared with a similarly 'recyclable' material such as stone, where the individual blocks retain form over a much longer period). This is particularly so when the conservation approach results in an alteration to the properties and values of the material, such as the use of consolidants which make the material 'unbreathable'.

The use of traditional materials and techniques for conservation work may be associated with a new interest in retaining and using skills associated with earthen architecture. However, the nature of the conservation intervention may influence the skills associated with the material in 'living contexts'. In all of these instances the materials and techniques used and the condition of the conservation interventions transmit subtle messages. For visitors to a site or building the utilisation of inappropriate conservation approaches or the poor performance of the conservation approach adds to the negative values and perceptions of earthen architecture. In other instances people who see material being used in conservation work might transfer the technology to their own domestic buildings. For example, the use of cement-based renders for consolidation or reconstruction is particularly problematic, this could generate the notion that it is acceptable not to maintain, and on the other that the utilisation of harder cement-based materials is appropriate for earthen architecture. Given the problems associated with these conservation approaches, transferring these approaches between different contexts of engagement with earthen architecture (from conservation of an archaeological site or historic building to maintenance and construction in a living context) may limit the life span of earthen architecture, and serve to generate and reinforce the negative view of the material.

At a broader scale, the values associated with earthen architecture are influenced by changes in the way traditional skills are acquired, and the value placed on traditional skills and craftsmanship. These skills (and contexts of acquiring these skills) have altered in the 20th century, influenced by the globalisation of a modern building style and materials concerned with the assertion of modernity (see *Chapter 3 & 4*). The impact of these changing perceptions of traditional construction and craftsmanship has resulted in a decline, and alteration, in the skills associated with earthen architecture. As a result, skills essential for maintenance and new construction may be lacking and altered, for example in Uzbekistan the traditional shapes and sizes of mudbricks have altered to share similarities with cement breezeblocks (observations in Samarkand, 2004). Again, the decline in traditional skills can reinforce negative associations of earthen architecture; badly built structures may be more liable to collapse, whilst poorly maintained structures will have a reduced lifespan.

The alteration in the type and transmission of skills associated with earthen architecture also impacts upon conservation approaches. On a number of sites conservation specialists have carried out training programmes particularly concerned with improving the properties and durability of the basic earth building materials and technologies. This can raise issues associated with the suitability of the taught techniques if they are not seen within their local context, for example 'specialist' knowledge concerning the inclusions within mudbricks and earthen plasters can contrast with local knowledge and practice. These issues are particularly problematic when conservators are aware that different inclusions (or different methods of working the inclusions such as finer chopped and rotted straw) will improve the performance of the conservation material, or if archaeological and historical research indicates different inclusions were used in the past that are no longer used in contemporary earth building contexts. For example, at Merv the work undertaken for the repair and maintenance of the domes on the Mausoleum of Ibn Zeid was problematic as contemporary earthen building practice produced an earth plaster that performed poorly, did not reflect the historical and oral references to different inclusions within the earthen mix, and did not reflect a 'scientific' understanding of the performance of the earthen materials (*Appendix 4 & 6*). In these instances the decision to change the material and techniques used for earth construction and maintenance for conservation in order to reflect the better understanding of the materials performance or historic use poses issues associated with the materials authenticity. In turn this can impact and alter the values associated with earthen architecture as rooted distinctively within community and locality.

Conservation approaches can threaten the values associated with earthen architecture as an autonomous and locally distinctive building material. For example, conservation approaches that obscure locally distinctive forms of earthen architecture (such as placed earth) behind newly manufactured mudbricks impact the understanding and interpretation of a structure, whilst reducing its local distinctiveness. Similarly conservation solutions involving the replastering of the surface of the earthen walls, regardless of whether or not this had occurred on the original (placed earth walls, for example, were often not plastered), or the replastering of excavated archaeological sections and baulks, significantly impact upon the understanding and interpretation of a structure.

I have argued that earthen architecture is environmentally aware and friendly (*i.e.* suited to and adapted by its environment) (see *Chapter 4*). Unfortunately, a number of conservation approaches can alter the environmental values associated with earthen architecture. For example, some conservation approaches may threaten the values associated with earthen architecture as environmentally responsive, such as seeking to consolidate exposed earthen architecture without a protective roof (as this suggests the building material can survive without being adapted to its environment). Other approaches may challenge the value of earthen architecture as environmentally friendly, such as the utilisation of materials whose manufacture poses environmental problems, such as geotextiles, consolidants (products of the petrochemical industry), or cement (the manufacture of which contributes to 12% of global CO^2 emissions). In other instances conservation approaches can threaten the sites environmental setting, such as the alteration in the environment through the installation of over-effective drainage. A number of the conservation approaches assessed through this research have impacted the environmental and aesthetic values associated with earthen architecture, through the alteration of the site, the form of the structure, or the site setting (such as the installation of shelters, utilisation of reconstruction, and to some extent, backfilling of archaeological trenches, see *Chapter 8*). Often this is the fault of the context of the intervention (such as financial limitations) that poses particular problems when planning for holistic site-wide approaches for the conservation and management of earthen architecture.

In conclusion, the values associated with earthen architecture impact approaches to its conservation and management. Time and again through the period of study, the conservation approaches recorded and studied reinforced the negative view of the material. It is still difficult to find an approach to the conservation and management of earthen architecture that finds a balance between improved condition, conservation theory, the retention of the values associated with earthen architecture and the wider impact of the intervention. I would argue this could only be achieved through holistic approaches to conservation and management that see the retention and exploration of the values associated with earthen architecture as key to success (as advocated through the framework set out below).

Sustainability

Chapters 1 & 7 identified sustainability as a contemporary value linked to the environment, which recognises the importance of the past, *and* how current use may pose tensions for the future of the resource. The broad definition of sustainability encompasses the economic sustainability of the conservation approaches (which is much easier to assess and quantify), the physical and environmental impact of the conservation approaches, alongside broader, more holistic notions concerned with equality between and within generations. These notions of sustainability impact and are impacted by the approaches and materials used for the conservation of earthen architecture.

The sustainability of some conservation approaches on some sites is limited given the sheer scale of earthen architecture to be retained and preserved, where an approach may work well on a small scale, but its application over a wider landscape or structure is problematic (such as undercutting repairs). With many conservation solutions for earthen architecture there is a tendency to see interventions as final 'one-off' solutions, after which there may be a decline in interest in the work carried out: this is problematic as all conservation work requires maintenance. This is symptomatic of the approaches to conservation through encapsulation and restoration recorded in Uzbekistan, where there has been no maintenance of the (often substantial) work carried out (*Chapter 9 & 14 & Appendix 6*). There is a need to see the conservation of earthen architecture as a long-term, holistic process for remedying the effects of erosion and deterioration. The factors that result in the loss of earthen architecture are both continual and interconnected (*Chapter 5*), even when a site or building is conserved erosion will continual and maintenance is essential.

Some conservation approaches lack sustainability as they result in further damage to the site or structure being conserved, such as the use of cement renders that pose problems in the long-term associated with increased deterioration and difficulties of removal (see practical effectiveness above). If we accept that few of the conservation solutions utilised for earthen architecture provide a long-term solution for the retention of the material then approaches that utilise monitoring and maintenance must be advocated as more sustainable.

The economic sustainability of many of the approaches to the conservation of earthen architecture is questionable (such as encapsulation/restoration work, and sheltering (*Chapter 9, 14 & Appendix 6*). In many instances funding and investment is sought for the initial conservation work, but long-term investment to assure monitoring and maintenance is problematic as this is associated with the context of the conservation work which is dependent on infrastructure and empowerment. This is particularly problematic as all of the conservation approaches require monitoring and maintenance. A lack of maintenance results in a limited lifespan for the conservation work, leading to deterioration which can often be quite rapid, depending on the local environment. The economic sustainability of conservation would be much better assured if funding bodies approved funding for documentation, monitoring and maintenance, rather than just the capital costs of the initial conservation work (for example the shelters at Çatalhöyük (Turkey), and reconstruction work in Central Asia (*Chapter 14*)). Often the limited funding for site conservation and management activities is supplemented through tourism, but there is a tension between increased funds and the increased rates of deterioration associated with higher visitor numbers (see *Chapter 5*). Similarly there is a broader environmental impact of some of the materials and techniques used for the conservation and management and this can be seen as contributing to patterns of climate change (see above).

I have argued sustainability is a contemporary value linked to the environment and I would argue many conservation approaches lack sustainability as they do not consider how broader environmental change may impact on them in the future. The effects of climate change in altering annual temperature and rainfall patterns are beginning to be recorded and felt around the world. According to the World Health Organisation 150,000 people are already dying every year as a result of the impacts of climate change, including droughts, floods and storms (WHO Report 2003). Future climate change models see impacts associated with more variable temperature and moisture regimes, with some regions becoming drier and colder whilst others will become wetter and warmer. This will impact on the natural and human environments, shifting populations and altering land use. Taking into account the impact of future climate change models on the factors causing deterioration to earthen architecture is an important next step in providing for sustainable approaches to the conservation and management of earthen architecture in archaeological contexts. Climatic change may have both positive and negative effects on the survival of earthen architecture. For example, within the study area one climate change model for Central Asia predicts higher winter temperatures in Nepal and Afghanistan, which will reduce the amount of spring meltwater reaching the Murghab, Amu Darya and Syr Darya river systems that feed Turkmenistan and Uzbekistan (Lynas 2004). As a result these areas may become drier with an increase in desertification which may improve the survival of earthen architecture. However the impact on the human populations within the area would be much more problematic, potentially leading to the abandonment of large areas of settlement. This extreme scenario means that any future conservation and management strategy that saw maintenance through the employment of local labour would be inappropriate (as there would be no local labour force); rather the monuments and sites would be 'abandoned' and continue to actively erode but with a higher degree of survival because the erosion factors associated with excess moisture would be reduced.

Within this bigger environmental context there has been little consideration of the association between the survival of earthen architecture and changing agricultural practices, salination and contamination by agrichemical residues. There may be problems with the use of materials for conservation that have been contaminated by chemical fertiliser (the possible relationship between the presence of nitrates and the premature deterioration of materials). It would seem appropriate to apply a similar warning to the earthen materials (and additives to the earthen materials) used for maintenance and conservation. Future research on environmental contamination may better understand and indicate further problems (see below).

A holistic view of sustainability recognises the importance of the past, *and* how current use may pose tensions for the future of the resource, envisaging a complex and holistic relationship between people, the environment, and the material remains of the past, for the present and for the future. In conclusion, whatever conservation and management approach is adopted for earthen architecture there is a significant impact on the past, present and future of the resource. Many conservation interventions impact on the future of the resource (impacting its shape, form and values), whilst others impact upon the local, regional and global environment within which the resource is placed. The sustainability of earthen architecture is threatened by conservation and management-planning activities that are not placed within the local political, environmental and social context. As seen from the global and regional studies, the conservation approaches often lack the holistic assessment of both local and wider impacts (such as environmental, social, aesthetic and interpretive effects), alongside the balance between the practical effectiveness and retention of values within the context provided by conservation theory. It is this type of broad assessment of impact and balance that would better assure the sustainability of conservation approaches for earthen architecture (see below).

Summary

This research has shown howa variety of approaches to the conservation of earthen architecture developed through the later half of the 20th century. Most of these approaches can be criticised in some way; limited practical effectiveness, interpretation (or misinterpretation) of current conservation theory, negative impact upon the values of earthen architecture, or threatening the sustainability of the resource. Most of the approaches, however, can be understood given the contextual basis of the intervention and the interplay between conservation and contemporary society. Site managers, archaeologists and conservation professionals all around the world are trying to meet the needs and demands of contemporary society, often with misconceptions and negative perceptions of the material they are working on, often without the information of what has and has not worked well in other contexts, often with an interpretation (or misinterpretation) of current conservation theory, and often within tense political and economic contexts. Given the context within which conservation activities are undertaken it is understandable why certain approaches have (and have not) been adopted for earthen architecture.

A transferable intellectual framework for earthen architecture on archaeological sites

One aim of this research was to establish a transferable intellectual framework to assist in the conservation and management decision-making process for earthen architecture on archaeological sites. As this study has shown, no matter what approach is adopted for the conservation and management of earthen architecture there are positive and negative impacts in relation to conservation practice, theory, the values of earthen architecture, and sustainability. For the 21st century I would argue that we need a new proactive and empowering framework for the conservation of earthen architecture. This framework is developed from an awareness of the physical properties and values associated with earthen architecture and is, therefore, pragmatic, flexible to the needs of a changing environment (where erosion and deterioration in different contexts may be occurring more or less rapidly), and aware of the contextual basis of our interaction with the archaeological and historic environment.

The purpose of this framework is to provide a group of ideas and concepts within which we can operate alongside the conservation actions I consider appropriate for earthen architecture. The intellectual framework is concerned with both the broad concepts I consider essential for consideration within the conservation decision-making process, and with the conservation actions I consider appropriate for earthen architecture. The framework is not prescriptive, rather it emphasises that conservation approaches and actions are contextually based and derived. As such the framework proposes a set of transferable broad concepts and actions concerned with the conservation and management of earthen architecture. It also seeks to develop concepts and actions that can be transferred between and within different contexts of interaction with earthen architecture.

The framework is envisaged as enabling future decision makers to have a basis upon which decisions can be based, concerned with both the practical issues of 'what we do' (using the multitude of different approaches, techniques and materials identified in *Chapters 8-16*) and the understanding of 'why we do it' within the context of conservation and heritage theory.

In this respect the broad concepts and actions proposed by the framework fit within contemporary approaches to conservation and management planning proposed by the 1999 Burra Charter, and other contemporary conservation planning models discussed in *Chapter 7* (Avrami *et al* 2000; Clark 1999, 2002; Demas 2002; Mason 2002; Mason and Avrami 2002). These models define contemporary approaches to an iterative conservation planning process broadly concerned with identification, understanding significance, developing policy, and managing. In addition, the 1994 Nara Document on Authenticity widened the concept of authenticity, linking it to form and design, materials and substance, use and function, traditions and techniques, location and setting, spirit and feeling. This framework for earthen architecture could sit alongside these contemporary approaches to conservation and management planning assisting in the identification of heritage assets, understanding significance, developing policy (informed by the broad concepts), and managing (through the actions appropriate for earthen architecture). The transferable framework for earthen architecture proposes broad concepts that are concerned with our approaches to the site and the material, whilst the actions underpin the vital importance of identifying, documenting and understanding the resource (the site and the materials it comprises). It is in the recommendation of specific actions that this transferable framework moves forward from the already established approaches to conservation and management planning.

Framework for earthen architecture - broad concepts.

Sustainability

By planning carefully and looking for a sustainable approach to the excavation and conservation of earthen architecture the notion of this material as 'unconservable' can be challenged. In doing so the very notion of what is and what is not 'conservation' in relation to earthen architecture is also challenged. I would argue we should be looking for a new terminology – one of the sustainability of earthen architecture. By using this term with its implicit notion of equality between past, present and future generations, we can embrace and interact with earthen architecture in all contexts.

Within the framework, sustainability is also concerned with the economic sustainability of conservation activities on a site, and this is concerned with aspects such as tourism and income generation (to enable work to be carried out, and enable work to be monitored and maintained). In other instances sustainable approaches to cultural heritage may place a much greater emphasis on the local capacity to carry out planned works, querying conservation approaches that rely on outsiders, or expensive imported conservation materials. Key to the notion of sustainability is the involvement of people and the realisation of the economic impacts of resource use, through the employment of local populations, and the gathering of local knowledge. Ensuring the sustainability of conservation approaches raises issues associated with how current activities or planned activities will impact the current and future retention of the resource, whilst balancing the needs and expectations of local communities in the present and future.

The conservation decision-making activities should be placed within a sustainable development framework, underpinning this notion is the important role of equality and fairness, concerned with the sharing of knowledge, expertise, and wealth.

Archaeology and conservation; conservation and archaeology

Understanding the importance of the archaeology of conservation and the conservation impact of archaeology is particularly important for earthen architecture. As such we must be aware that earthen architecture is constantly eroding and subject to archaeological formation and deformation. Understanding the constant nature of erosion phenomena (see *Chapter 5*) corrects the misunderstanding of unexcavated archaeological sites as being 'static' (which if left untouched, would result in the 'permanent' retention of the buried archaeological deposits (as argued by Carter and Pagliero 1966, 67)). This misunderstanding has contributed to a generally negative attitude towards excavation from conservators, and has resulted in most conservation activities on archaeological sites being restricted to an intervention-by-intervention approach rather than a holistic approach to the whole site (as the excavated area is perceived as the only actively eroding zone).

However, any intervention whatsoever within the eroded and eroding archaeological deposits disrupts and alters the natural and active patterns of erosion, and these disruptions can result in both positive and negative change. In the 21st century archaeologists need to be much more aware of the impact of their activities on conservation and there is a need to consider techniques of archaeological excavation that are more appropriate and sensitive to the needs imposed by the characteristics and deterioration phenomena that effect earthen architecture, for example positioning trenches in locations less likely to impact natural drainage and run-off patterns, or only undertaking excavation in an environment moderated by a shelter.

Similarly, conservators need to be much more aware of the archaeological impact of their actions as there is an archaeological impact from many of the conservation solutions utilised for earthen architecture, such as the quarrying of new earthen materials for use in conservation or the below ground impact of drainage works and undercut repairs. Similarly those involved in new construction need to be aware of the archaeological implications of material acquisition and quarrying.

In a broader context archaeologists and conservators need to be part of a team, aware of the impact of their work on part of a site over another, and to understand that archaeological and conservation activities are contextually derived with a multitude of suitable solutions. Archaeologists and conservators must be much more aware of both the universality and local distinctiveness of earthen architecture in the past in order the better identify, document and understand the archaeology and conservation of earthen architecture.

Compromise

Conserving the values of earthen architecture may mean compromise. Within the context of contemporary conservation and management planning disciplines the need to balance different needs and aspirations is recognised. Often this approach has focussed on balancing the needs of different stakeholder groups with diverse and conflicting views of a site or structure. Extending this compromise in relation to the conservation of earthen architecture should also consider that sometimes the sustainability of the resource is only assured by compromising on some of the practical and theoretical aspects of conservation and management (particularly where balance is needed between other aspirations such as poverty relief and use).

Consistently and holistically

Conservation and management activities must be planned holistically. This is concerned with looking at the *entire* site (not intervention-by-intervention), and *all* of the erosion and deterioration factors. Approaches, materials and techniques should also be consistent and self-contained over a monument or site to limit the negative impact of conservation interventions on site understanding and interpretation (within a framework that can reflect changing knowledge and/or the results of monitoring through time).

Holistic planning is essential as understanding the different stages of loss is important in assessing the suitability or otherwise of future conservation and management interventions for earthen architecture. For example, in the early stages of deterioration, if buildings still have roofs, interventions that seek to retain those roofs through repair, and the placing of the structure in a management context in which it will receive maintenance, and regular maintenance checks (possibly through the adaptive re-use of the structure) is a sustainable approach to the retention of the resource. In contrast efforts taken to retain a very eroded and eroding wall stump surviving on top of an archaeological mound, may rather seek to document the current condition of the resource and place the resource in a management context in which the whole site will be protected from future development, looting and damage from birds, insects and burrowing animals. This approach would retain the whole site, rather than the wall stump and seek to retain the future information values associated with the entire resource rather than just the physical remains of the wall stump.

Context

An emphasis should be placed on understanding the contextual basis of attitudes to, and values associated with earthen architecture alongside the contextual basis of conservation approaches. As this research has shown, the different values associated with earthen architecture are rooted in locality, and these values impact upon the methods of retention.

Climate change

Environmental concerns should be brought into the mainstream of planning for the sustainability of earthen architecture. Future conservation and management planning should take into account and pragmatically plan conservation activities in relation to human induced climate change. Looking into the 21st century the approaches appropriate for earthen architecture may come to reflect the local impact of alterations in global climate. This means that in some locations greater intervention may be appropriate (those with increased and more erratic rainfall, and alteration in groundwater), whilst in other locations (those with increased aridity) less intervention may be appropriate for assuring the sustainability of earthen architecture. In other instances the environmental impact and ecological assessment of proposed conservation work (such as the materials utilised, or travel by a specialist to a site) should be considered as part of planning for conservation activities on site (through the use of ecological footprint analysis (Chambers *et al* 2000) (or the further development of this process specifically for cultural heritage contexts)).

Locality

The decision-making process must be rooted with people locally in order to reflect the contextual basis of conservation approaches alongside the values associated with earthen architecture. Rooting the process within the locality makes explicit the connection between earthen architecture and people. As I have argued in *Chapter 6*, people provide the mechanism which enables the resources to be retained and sustained, there is an explicit connection between the causes of change (as manifested in the loss of people, population change or shift) and the onset of threats and loss to earthen architecture. It is vital to understand and make explicit the connection between different localities, contexts, materials and different techniques with which people retain and sustain the values associated with the earthen architecture. In this respect rooting the decision-making process locally and within a community draws reference to the 2004 INTACH Charter in India which is concerned with identifying a sustainable interpretation of contemporary heritage theory in order to retain traditional craft skills, preserve cultural diversity and local distinctiveness, and to improve social and economic conditions.

Flagship projects

There is scope to examine the use of earthen (and other traditional) building materials and techniques in flagship projects for sites and structures, such as interpretation centres, museums and artefact stores. The careful utilisation of the sort of technology and expertise developed in Western Europe for the construction of modern earth structures offers the potential to retain traditional skills in other contexts around the world, particular where thought is given to integrating 'new technologies' alongside the retention of locally distinctive forms of earthen architecture. This could challenge negative values associated with earthen architecture and use this new interest and new perception of earthen architecture (for the generation of a 'culture of acceptance' *pers comm.* Tom Morton) as a means to retain earth-building skills (for maintenance, conservation and new construction). Such places could be constructed with a small ecological footprint, utilising the passive thermal and moisture regulation of earthen building materials to regulate the interior climate without recourse to climate control. Through the re-valorisation and retention of earthen building skills the local distinctiveness of an area can also be retained. These sorts of projects underline the important connection between past, present and future.

Similar projects concerned primarily with education or the provision of low-cost sustainable housing have been utilised throughout the world, such as the DRUK White Lotus School in Ladakh, (Architecture for Humanity 2006), on archaeological sites at the eco-centre at Gordion (Turkey) (Summer 2003), and Dakhleh Oasis (Egypt) project dig-house (Schijns forthcoming).

Preventative conservation

Preventative conservation is concerned with identifying all of the factors resulting in loss across the entire site. This is in order to counteract some of the problems associated with conservation solutions that sought to remedy one cause of erosion and deterioration and have (inadvertently) contributed to further erosion and deterioration. A preventative conservation approach reflects the concerns of current conservation theory, and is also a suitable and sustainable framework for earthen architecture. In this context preventative conservation is concerned with identifying the causes of erosion and deterioration, and taking action to minimise or eliminate damage, such as through the creation of a stable environment through sheltering.

Precautionary approaches

Implicit within the notion of sustainability is the concept of adopting a precautionary approach, this may mean that rather than opting for large, risky conservation approaches, smaller, less experimental approaches are adopted for the conservation of earthen architecture. Such would be the pragmatic decision to utilise traditional earthen mortars and plaster for surface treatment rather than experimental consolidants; in other instances approaches used as standard practice in other disciplines (such as revegetation and slope stabilisation in geotechnical engineering) may be appropriate, as the risk and problems have already been researched and documented elsewhere.

Traditional, indigenous knowledge and know how – the intangible heritage

Sustainable approaches for the conservation of earthen architecture see a much more interdependent role between the intangible and tangible aspects of the earthen architecture legacy. The intangible heritage of earthen architecture is as important as the physical remains of the past, for example local practice and knowledge concerned with the beneficial role of additives to the basic earthen mix, and techniques of maintenance and construction. This intangible heritage presents authentic approaches to earthen architecture and fits within contemporary approaches to conservation recommended within the 1994 Nara Document on Authenticity and 2004 INTACH Charter in India. As such, approaches to the conservation and management of earthen architecture should place particular importance on the identification, documentation and (where appropriate) utilisation of traditional, indigenous knowledge of earthen building materials and techniques.

Framework for earthen architecture – future actions.

Documentation

Documentation is highlighted within contemporary approaches to conservation and management planning proposed by the 1999 Burra Charter, and other contemporary conservation planning models discussed in *Chapter 7*. For earthen architecture in particular emphasis in the future should be placed on documentation, concerned with (1) archaeological documentation of historic materials, (2) documentation of conservation interventions, and (3) ethnographic documentation of contemporary practice and local distinctiveness including the identification of earthen architecture types, local skills, and intangible, indigenous knowledge.

I would argue that by understanding and accepting documentation as a valid approach to conservation, the notion of earth as 'unconservable' is challenged. This emphasises that sometimes we are concerned with retaining the values of earthen architecture, rather than the physical remains. This approach to conservation emphasises the notion of 'preservation by record' and embraces a much wider and complete scale of documentation than that used currently in many developer-led archaeological, and conservation contexts, where often the scale of documentation and sampling of the material undertaken is too little to understand the scale, depth and complexity of the archaeological record. For those contexts where documentation is accepted as a valid conservation approach I would argue that the data collection should incorporate materials analysis, written, photographic and drawn records, alongside more complex 3-D recording of current condition (which if undertaken using a 3-D scanner is fully repeatable and thus able to monitor change over time). Implicit within all of these recommendations is access to a useable and worthwhile dataset, enabling reference to be made not just to the different approaches but also to the materials and techniques utilised.

'Doing nothing'

Earthen architecture sites and structures *can* survive (and often survive better) without large-scale intervention. In some instances non-intervention on a site or structure may therefore be an appropriate solution. As shown through the study of the historic photographs at Merv the deterioration and erosion of earthen architecture is a non-linear phenomenon, the structures suffer from gradual attrition which is occasionally punctuated by episodes of greater loss (*Chapter 5* & *Appendix 5*). In these contexts the erosion and subsequent formation and deformation of earthen archaeological deposits can retain the values associated with a place in a 'transformed' state.

If taken deliberately within a management context (where it is protected from development, etc) and documented in detail, a policy of non-intervention can be both logical and realistic. In this context 'doing nothing' is sustainable as this does not necessarily impact the future decision making process, and does not necessarily limit the resources future potential. Making reference to the 19th century conservation debate John Ruskin used the formation process undergone by deposits of earthen architecture in his anti-restoration argument highlighting the extent of preservation in tells, implicit within this argument is that by 'doing nothing' the information value of a site can be retained:

"Do not let us talk then of restoration. The thing is a Lie from beginning to end ....the old building is destroyed, and that more total and mercilessly than if it had sunk into a heap of dust, or melted into a mass of clay: more has been gleaned out of desolated Nineveh than ever will be out of re-built Milan." (Ruskin, 1880: 196)

In other instances materials and techniques for conservation activities may work well, but they may also be expensive, they may be reliant on outside specialists for their use, and the manufacture may impact the environment; so rather than use the materials and techniques the decision may be taken on a site to compromise and 'do nothing'.

In this framework the 'do nothing' response is therefore not passive, but is concerned with assessing the past, present and future of the resource. Sometimes 'doing nothing' does actually mean doing some things, for example the monitoring of a site or monument in which such as approach is adopted may lead to the revision of this approach, perhaps with preventative conservation solutions adopted in the future.

Monitoring

Monitoring is a significant aspect of planning for the management and conservation of earthen architecture. Monitoring is vital for (1) establishing current condition, (2) assessing conservation work carried out, and (3) for those sites or monuments in which 'doing nothing' is the management option. Monitoring may indicate when maintenance is required, or for those sites in which 'doing nothing' is a management option, monitoring is significant in understanding the erosion and deterioration process, and may lead to a revision of the conservation and management approach adopted. Any system of monitoring is underpinned by the skills, techniques and capacity for documentation activities (see above).

Maintenance

Maintenance assists in assuring the sustainability for the resource and community. Maintenance can be a significant and suitable approach for the conservation of earthen architecture (in both living contexts and archaeological contexts) based on the physical properties and values associated with earthen architecture. Rather than viewing maintenance as a destructive practice (such as at Çatalhöyük (Turkey) (*Appendix 6*)) maintenance should be seen as both a relevant component of contemporary conservation practice which retains intangible heritage, and an important method of asserting and reflecting the values of earthen architecture.

In addition, this research has shown that too often the limited interpretation of 'conservation' is seen as a one-off solution, after which the work is left unmonitored and unmaintained. However maintenance is also required for all of the conservation work that is carried out on a site (and if conservation work is carried out with earthen materials some of the materials should be left in storage to easily carry out required maintenance). The maintenance activities carried out on earthen architecture should be informed by monitoring and documentation.

Using the framework

Within this aspirational framework for the sustainability of earthen architecture my approach to the conservation of earthen architecture is pragmatic and contextually dependent. In the first instance I am concerned with understanding the values and associations of earthen architecture in the context within which I am working. I am concerned with documenting earthen architecture in its current condition, and then I'm happy to think about, and be flexible according to the context I am working in, either documenting and doing nothing (leaving a record of the site, structure or earth building practice); or monitoring and maintaining a site or structure through simple management methods (such as rubbish and vegetation clearance, where appropriate, using traditional earthen materials for maintenance work, and where structures still have roofs trying hard to keep that protective roof); or undertaking more substantial interventions for a whole site or structure (such as backfilling, re-vegetation, sheltering, or restoration using appropriate materials). These solutions are not right for every site, but operating within the transferable intellectual framework developed through this research should enable the right decision to be made for the sustainable conservation and management of earthen architecture (see sketch 5).

Criticism of the proposed framework

Critics would perhaps comment that by focusing on the values associated with the material the transferable framework for earthen architecture proposed by this research is looking to the past, to the conservation debate of the 19th and 20th century, rather than to the current emphasis of the Burra Charter and the broad stakeholder and value-based management planning process. To answer those critics, I would query to what extent are the values of stakeholders actually taken into consideration when their views are widely opposed to those of 'conservation professionals'; and is the assignation of 'values' really holistic and participatory? The gap between the international/western conservation theory and indigenous principles and practices of conservation is not unique to earthen architecture (as the 2004 INTACH Charter in India shows).

Too often on the sites visited during the period of this study the conservation approaches were not holistic, and they are perhaps concerned more with perceptions of what 'conservation' is supposed to be, rather than the requirements of the sites, structures and locality. If archaeologists and conservation professionals are actually concerned with the understanding and the 'retention of the values associated with a place' then one of those values is the value associated not just with the 'place', but also with the materials from which the 'place' is comprised. For earthen architecture in particular, where vast, enormous archaeological sites and structures are formed of eroded and eroding earthen building materials the need for a holistic approach, considering the physical properties and values of the material is vital. I would argue that by stressing the physical properties and values of earthen architecture within the conservation and management planning process the proposed approach is rooted within current approaches to value-based management planning (see above).

In other respects further differentiated is required between the proposed framework for earthen architecture and the 1994 Nara Document on Authenticity. This document linked authenticity to the understanding of a broad range of heritage values, and sources of authenticity in form and design, materials and substance, use and function, traditions and techniques, location and setting, and spirit and feeling, and other internal and external factors (Article 13 Nara Document on Authenticity). The framework for earthen architecture is developed from the awareness of the contextual basis of our interaction with the archaeological and historic environment. In this respect aspects of the transferable framework for earthen architecture such as emphasis on context, locality and local knowledge (intangible heritage) are all within the spirit of the Nara Document. This is underpinned by the emphasis on maintenance as a future action to assist in assuring the sustainability for the resource and community both for living contexts and archaeological contexts.

What makes the transferable framework different to the Nara Document is the emphasis on the physical properties and values associated with a single material. As such by emphasising the uniqueness of earthen architecture this transferable framework can be used for earthen architecture in different contexts. By focusing on the material (as something that can we can touch and converse over) the framework can cut across social, cultural, gender and age boundaries with greater ease than more conceptual conservation ideas (such as authenticity despite the significant develops that resulted from Nara). As a result of focusing on the material this transferable framework also proposes concepts and specific practical actions in a manner which distinguishes this from the Nara Document. As is common to a majority of contemporary approaches to conservation theory this intellectual framework is pragmatic and flexible. However this framework emphasises the contextual basis of our interactions with the archaeological and historic environment alongside the physical needs of earthen architecture in our changing climate and environment.

Conservation solutions – difference and 'otherness'

The third aim of this research was to develop a wider understanding of 'difference' in approaches seen within conservation and heritage theory. The identification of what lies outside the observer's own cultural experience and the perception of difference, distance and otherness is a characteristic of the idea of 'us and them' identified by Edward Said as concerned with control, influence and the assertion of supremacy (Said, 1993 xvi). Furthermore, Gosden comments, "we in the West structure our thought around a series of polarities – being *vs.* nothing, man *vs.* woman, speech *vs.* writing – in which the second term is seen to be a negative, corrupt version of the first" (1994, 55).

Chapter 7 identified how the observation of difference has been used to critique approaches to conservation (for example, Lowenthal 1985; Stille 2002). These observations have been extended to envisage a dichotomous relationship between the international/western approaches to conservation, and approaches found elsewhere, for example Cleere identifies a restricted concern for archaeological and historical artefacts in, "less-developed societies" (1989, 6). A simple value-based dichotomy is envisaged: good conservation characterised by conservation theory *vs.* bad conservation characterised by approaches seen elsewhere. These observations of difference are based on the comparison at a global scale between contexts of use, maintenance and repair, and, contexts of abandonment, conservation and restoration.

The data concerned with earthen architecture collected through this research allows an understanding of difference and otherness in relation to a single class of material. The data collected and analysed by this research enable me to conclude that conservation activities are contextually dependent, and 'differences' in approaches result from the complex interplay between conservation and contemporary society. This shows that the physical properties and requirements of earthen architecture are just one factor that impacts upon the materials, techniques and approaches used for its conservation.

I would argue that 'difference' has often been observed and valued in relation to differences in approaches to the conservation of 'incomparable' materials, such as the conservation of stone and timber structures, which have very different physical properties, values and associations. As these different materials share different properties and characteristics there are different ways to approach their retention. As such the 'difference' observed in approaches to the retention of the material remains of the past, is associated with geographic and geological context (for example the types of material available for construction and the very variable types of erosion and deterioration). As a result I have used the

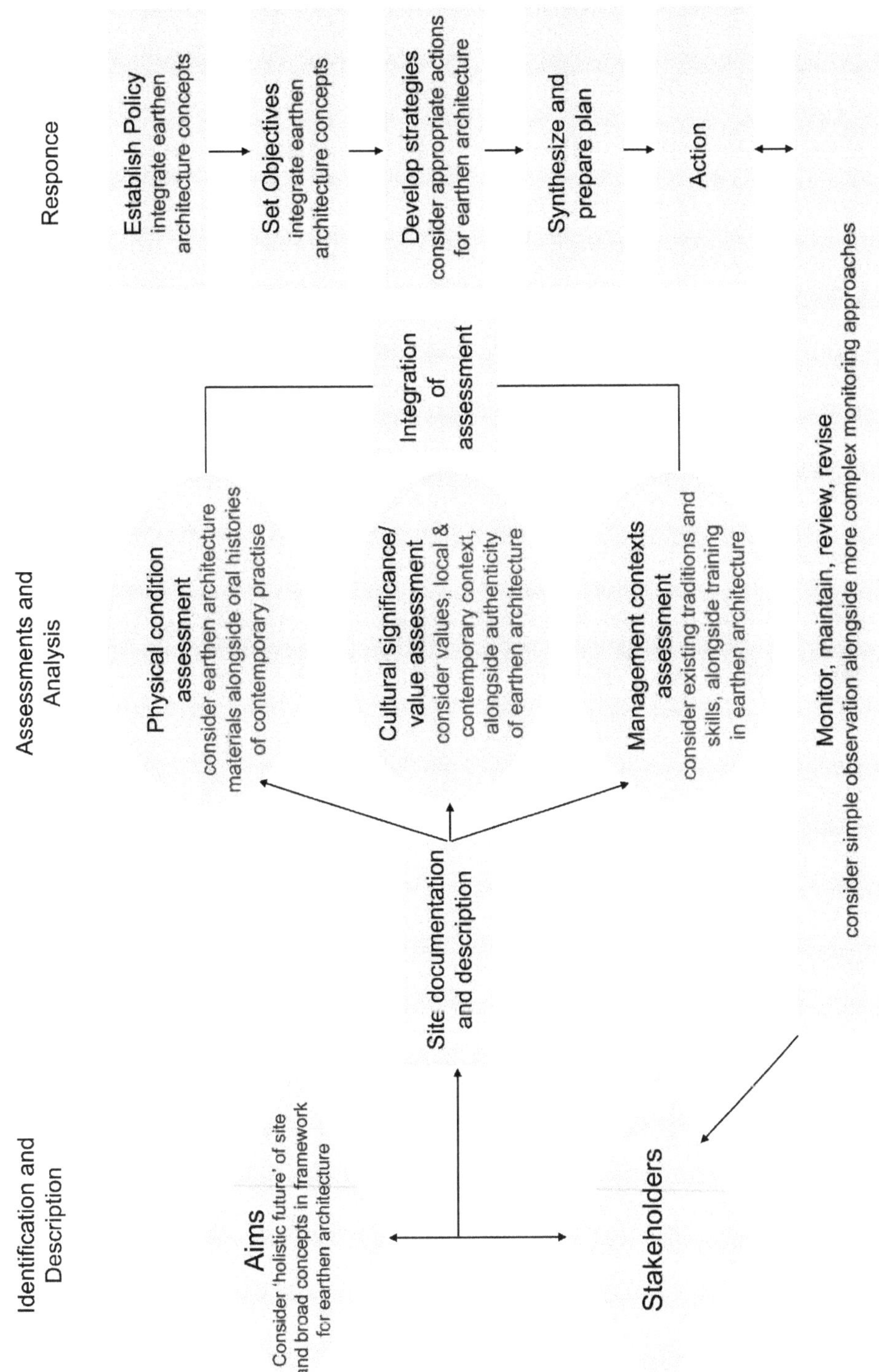

Sketch 5. A proposed management planning process adapted for earthen architecture (Diagram adapted from Mason 2002, 6), incorpaorating the future framework for earthen archutecture.

focus on a single material provided by this research to better understand the observation and valorisation of 'difference'

Difference observed in relation to earthen architecture

The simplistic observation of 'difference' has resulted in the assessment of certain 'good' and certain 'bad' approaches to conservation. In relation to the conservation and management of earthen architecture these differences can be characterised by the materials and techniques developed as a result of, and advocated by, conservation theory (chemical consolidation, retention of the visibility and phasing of a structure; and use of replacement chemical or engineered materials, backfilling and sheltering); contrasted with approaches in the majority world that use traditional earthen materials and techniques for maintenance of earthen architecture, and/or for restoration and reconstruction.

This research has shown the great variety and overwhelming diversity of approaches, materials and techniques appropriate for the conservation of earthen architecture. What this dataset shows is that even with a single broad class of material there is still phenomenal variation in the approaches, materials and techniques utilised for conservation. This indicates that the nature of variation and difference in approaches to conservation is associated not just with the physical properties of the material, but also with the context within which conservation activities occur.

Context

The dataset shows different approaches to conservation associated with different contexts. For example, earthen architecture in living contexts is associated with maintenance as one of the most effective solutions to retaining earthen architecture, so when the context of maintenance activities alters different responses may emerge. These 'different' responses to earthen architecture may be:

- Re-use - repair and renewal of the structure, either with the same function or a function that has significantly altered.
- Abandonment - associated with the subsequent formation and deformation of earthen archaeological deposits, and potential identification and recording.
- Retention - conservation of the original structure (utilising the different conservation approaches); or the symbolic retention of the place within re-used building materials; or the retention of the values associated with a place.

These different responses to the material are determined by context (in the broadest sense comprising physical, temporal, spatial, social, economic, and political context). This broad context influences the notion of appropriateness, determining what is and what is not assessed as suitable for the retention of the archaeological and historic environment. Some of the similarities in approaches to conservation are associated with particular contexts, for example, archaeological sites tend to have approaches adopted to them that will 'freeze them in time' (such backfilling, consolidation, and sheltering); whilst living contexts tend to be maintained to enable buildings and structures to remain in use, or be adapted for re-use.

In understanding the different approaches to conservation the context of the complex interaction between conservation and contemporary society cannot be over-emphasised. For example, the different contexts within which archaeological research was carried out in Soviet Central Asia, resulted in a legacy of open and abandoned trenches, this contrasts so completely with the context of current archaeological research at Çatalhöyük (Turkey), which has resulted in a legacy of consolidation and sheltering of excavated trenches. Similarly the encapsulation and reconstruction of the city walls of Yazd and Bam in Iran; Khiva, Bukhara and Shahrisabz in Uzbekistan, and Merv, has made these monuments very impressive and eye-catching, but they all share a certain visual similarity. The motivations for these approaches are associated with the economic and political context of cultural heritage, where impressive city walls can define and raise the profile of a city, asserting power and acting as iconic 'pulls' for visitors to sites. Similarly the very variable nature of conservation approaches is linked to economics, with great polarities in approaches associated with the wealthy and with the poor. Such is the case with the old town restoration and maintenance in Yazd, where those who are economically marginalised maintain utilising traditional materials and techniques, whilst those who are more affluent restore and maintain, utilising replacement materials such as cement. In understanding the approaches for the conservation and management of earthen architecture recorded and assessed in this research, the most important factor is the context of the activities, rather than the impact in relation to conservation theory, practical effectives, physical properties and values associated with earthen architecture.

'Official conservation' and the generation of 'otherness'

There are a number of patterns revealed by the dataset analysed in this research. It is interesting to note those sorts of conservation approaches recorded and documented within the conference proceedings and publications supported from international heritage bodies. These tend to be the conservation approaches that most reflect the ideas implicit within conservation theory. For example, within the study area the most frequently documented approaches are consolidation, backfilling, sheltering and restoration, whilst those approaches observed on site visits (and those with the most impact) tended to be encapsulation and various different degrees of reconstruction. If the observations from within the study area mimic and match those in other regions of the world then there is an obvious tension between those approaches to conservation that are recorded, documented (and published), and those approaches that are not. Through the selective recording and publication of the different approaches to the conservation of earthen architecture international heritage bodies generate and perpetuate the approved and 'official' approaches that reflect the ideas implicit within conservation theory.

This process is also associated with the geographical spread of approaches to conservation and management. The analysis and comparison of the geographical spread of the documented conservation approaches shows enormous variation in those locations in which conservation activities and research have occurred. This analysis shows that on the whole most 'conservation' occurs in Europe (although those countries most represented are USA, Italy, Peru, Iraq and Iran. This implies that these continents and countries are most concerned with 'official' approaches to conservation advocated through conservation theory. This is the result of both the country of origin of those undertaking research in the field, and the fact their individuals can afford to attend, present and publish their research at conferences. It may also be an effect of those individuals undertaking research being more likely to be aware and undertake conservation activities and research within the 'spirit' of conservation theory, both because they may be more aware of the available literature (as it is accessible in their own language), and because funding bodies and agents will support these types of conservation activities and research. These approaches and individuals are therefore much more likely to be published within 'official' discourses on conservation. The available literature can therefore be seen as reflecting the particular philosophical and funding requirements of national and international heritage bodies that are concerned with the approaches advocated through conservation theory.

To some extent the geographical distribution can be seen as adding to the perception of Europe and the west being better able to undertake and carry out 'conservation' work. Within this context it can be argued that the observation of difference and dichotomy in approaches to conservation is part of a self-supporting notion of a western-based 'conservation' specialist. For example, the body of literature documenting approaches to conservation does not consider the various different approaches to conservation that fall outside the requirements of conservation theory. Similarly the evidence shows that most 'conservation' occurs in Europe, and what 'conservation' is documented as being undertaken elsewhere is primarily by western-based practitioners. These tensions between what conservation approaches are and are not recorded can again be seen as further generating and re-enforcing the notion of difference and 'otherness' recorded in relation to approaches to conservation.

The comparison of the geographical spread of the documented approaches to the conservation of earthen architecture similarly shows the great variation and inequality of the types and locations of conservation research. Europe is the most represented of the continents, whilst to some this might indicate that Europe has the most examples of earthen archaeological and historic sites, we know that this is not the case. Indeed given the geographical spread and patterns of erosion and deterioration of earthen architecture, those continents with most archaeological and historic earthen sites are exactly those continents that have missed out on being included within the published conservation research. This again shows phenomenal global inequalities, illustrating that those countries and continents with the greatest landmass (and with the greatest diversity of living, as well as archaeological and historical evidence of earthen architecture) have been excluded from the published research - this situation maps other patterns of global inequality.

The observation and valorisation of differences can be seen as another aspect of generating the perception of the 'conservation specialist' employed to advise and recommend different approaches to conservation. In some instances this is useful, specialists have the time and finance to research the materials and techniques appropriate for the different conservation approaches (in the study area perhaps best shown by the different methods of backfilling and repairing undercut walls at Merv). However the exporting of conservation techniques that have a proven practical effectiveness is very different to exporting conservation techniques that fulfil the notions of conservation theory (often without considering the local context of the interventions). In this respect I would argue that it is entirely suitable to publicise and advise techniques of conservation that have a proven practical effectiveness, however publicising and advising techniques for conservation solely because they fulfil the requirements of conservation theory whilst ignoring the wider context of the relationship between conservation and contemporary society would seem inappropriate. In some respects these observations sit alongside contemporary approaches to conservation indicated by the 2004 INTACH Charter in India, which bridges the gap between the international/western conservation theory and indigenous principles and practices of conservation.

It is problematic that most documented 'conservation' research and activities for earthen architecture, has been concerned with specific practical applications rather than the broader picture. As this research has shown approaches to the archaeological and historic environment are related not just to conservation theory, but also to the relationship between conservation and contemporary society, understanding that the approach, materials and techniques of conservation are contextually derived and dependent.

In understanding the different approaches to conservation this research has made clear there are differences in assessing different approaches to conservation in relation to practical effectiveness, and in relation to conservation theory. Arguably it is the practical effectiveness of a particular approach that is a more worthwhile assessment rather than the valorisation of approaches in relation to conservation theory. Through the particular concern with approaches advocated through conservation theory national and international heritage bodies can be seen as not only generating the idea of difference and dichotomy in approaches to conservation, but can also help to generate and re-enforce the negative perception of earthen architecture as an 'unconservable' material. This is because the reality of fulfilling the requirements of conservation theory are contrasted by the physical properties and values associated with earthen architecture. The very fact that the majority of the undocumented approaches utilised for the conservation of earthen architecture do not reflect conservation theory should not exclude them from discourses on the appropriateness or otherwise of use, particularly where they have a proven practical effectiveness given the context of the intervention.

Conclusion

The conclusions address three different areas of concern. In the first instance I address the research questions posed in *Chapter 1*, I then highlight the potential future research concerned with the conservation and management of earthen architecture, and finally I reflect more personally on this research within my final discussion.

Research questions

To conclude this research I return to the research questions established in *Chapter 1* and summarise how this research has addressed these.

- *Is current conservation theory applicable to earthen architecture?*

Yes, current conservation theory is applicable to earthen architecture, and is one of the aspects that should be used within the conservation and management decision-making process. However, sometimes there are problems with the application of conservation theory to earthen architecture (such as with notions of reversibility and visibility), and sometimes the notions implicit within conservation theory (such as 'conserve as found') do not sit easily with the physical properties and values associated with earthen architecture.

- *Can a transferable intellectual framework for earthen architecture be established?*

Yes, but rather than prescriptive recommendations of what and what should not be done for the conservation and management of earthen architecture I have reached the conclusion that the future sustainability of earthen architecture will be better assured if we base the decision-making process on a group of broad concepts and actions that reflect the contextual basis of conservation interventions.

- *Are approaches to conservation dependent on temporal and spatial contexts?*

Very much so - the dataset collected for this research shows that different sites in different locations have very different conservation and management approaches. Similarly approaches to the conservation and management on a single site shift and change through time. This research shows that these differences do not follow a pattern (in the past interpreted as a simple dichotomy in conservation approaches, often between east and west) but rather reflect the contextual base of conservation and management activities.

- *Can contexts of use, maintenance and repair, and contexts of abandonment, conservation and restoration, be comparable? How do these affect approaches to the historic and archaeological fabric?*

Yes, contexts of use, maintenance and repair, and contexts of abandonment conservation and restoration of earthen architecture can be compared. Again the study and comparison of these different contexts shows an overwhelming diversity of approaches to earthen architecture. These different contexts of interaction with earthen architecture are generally associated with different types of approaches to its retention, and approaches deemed suitable for one context may not be appropriate for use in another context.

- *Can conservation interventions be assessed within their context as a means to better understand our approaches to the historic and archaeological fabric?*

Yes, but this is complex and difficult. Often the context within which the decision-making process occurs is very complicated and is determined not just by the physical need to retain the material remains of the past but also by the complex social, cultural, economic and political context within which our interaction with the archaeological and historic environment occurs.

- *Are differences observed in approaches to conservation based on the comparison of materials with widely different physical properties?*

Yes and no. I would argue that the observation of difference in approaches to the conservation and management of the historic and archaeological environment is to some extent based on the comparison of materials with widely different physical properties. This means that sometimes it has been very easy to observe difference in approaches to the conservation of historic buildings and archaeological sites comprised of different types of materials. However, this is only one aspect of the observation of difference. The dataset and analysis undertaken in this research illustrates the overwhelming diversity of approaches, materials and techniques utilised for the conservation and management of a single broad class of material that shares similar physical properties. My research shows that differences observed in approaches to conservation and management are determined by the physical properties and values associated with the material alongside the context within which the decision-making process occurs.

- *Are differences observed in approaches to conservation based on the assumption that what is advocated by current conservation theory actually impacts conservation practice?*

To some extent - often it is easier to observe approaches to conservation as being a product of the use (or misuse) of conservation theory. It is too easy to assess and observe approaches to conservation only from the basis of difference between what conservation theory advocates and what occurs in practice. This is because the assessment and understanding of the contextual basis of conservation approaches is complex and difficult. This is problematic, as this research has shown that often it is the context of the conservation intervention that is most influential in determining the approach, materials and techniques used rather than conservation theory. In this respect it is perhaps self-evident that conservation theory is just one of the factors that should be used in assessing the suitability of the conservation approaches, materials and techniques for earthen architecture.

Future research

Significant areas are worthy of further research concerned with the conservation and management of earthen architecture, and broader issues concerned with conservation and heritage theory. These include:

- Research concerned with impacts of climate change and earthen architecture, globally and within the study area.
- A better understanding of the political, social and economic context of archaeology, conservation and contemporary construction of earthen architecture.
- Developing a set of tools for the better documentation of the materials and techniques utilised for the different conservation approaches for earthen architecture, concerned particularly with assessing and measuring the sustainability of earthen architecture and its conservation and management in different contexts.
- Use of earthen architecture in high status projects, including research on the uses of earthen architecture as passive environmental regulation for museum and archaeological stores.
- The collation of information concerned with the development through time and regional variation in forms of earth construction, in order to map and understand the temporal and regional variation of earthen architecture. .
- The development of protection and conservation systems for traditional knowledge and intangible heritage.
- Raising awareness of the threats associated with the loss of the tangible and intangible heritage of earthen architecture in the 21st centuries emerging markets.

Final discussion

The speed with which earthen buildings erode is dependent on the type of construction, alongside the context and environment within which they are located. Earthen architecture poses particular problems as it may erode quicker and leave less trace in most environments when compared with other building materials. The physical properties and values associated with earthen architecture may not wholly comply with current conservation theory. Too often criticism of approaches to conservation has been based on the assessment of work in relation to conservation theory, rather than the assessment of practical effectiveness given the physical properties and values associated with earthen architecture balanced against the context for successful practical applications provided by conservation theory.

This research was concerned not just with the assessment of the conservation and management of earthen architecture, but also to record and understand the notions of 'difference' in approaches to conservation and management through the investigation of approaches to a single, broad class of material. As such this research demonstrates the manner in which conservation and management interventions change is dependent on context, determined by the complex interplay between conservation, heritage and contemporary society.

By showing both the global nature of earthen architecture alongside the regional and local distinctiveness of use and associations of the material this research has explored notions of 'otherness' and value associated with the material. By its very nature, to many people in the temperate wet United Kingdom, earthen architecture is a material that embodies notions of 'otherness', and it is a material that we associate with other places and other people, and other people in other places in the past.

The future framework for earthen architecture developed through this research is concerned not just with understanding the contextual basis of approaches to the past, but also with the contextual basis of approaches to earthen architecture in the past, present and future. What is significant from the last century of research into the conservation and management of earthen architecture is the wealth of experience, wealth of techniques and approaches, and wealth of passion in this field of research.

Rather than criticise the divergent approaches to the conservation of earthen architecture I wanted this research to pull together, understand and synthesise that information - so that just on a personal level we had something to base our decisions on at Merv - but more than that to understand the interaction between conservation and contemporary society. I have been concerned with understanding and assessing the values of earthen architecture, as it is these values that define a theory within which we should be operating for the past (archaeological sites and historic building), present (contemporary society) and future (planning for new builds).

Earthen architecture has been used for the last ten millennia and is used universally. Through maintenance people are the most beneficial to earthen architecture, but they can also cause the most damage to it. Undertaking this research has enabled me to see that how we engage with a living, breathing material like earthen architecture, to assure its retention and future sustainability is a significant metaphor for humanity.

TABLES Refereed to in Chapters 7 – 16.

BACKFILLING

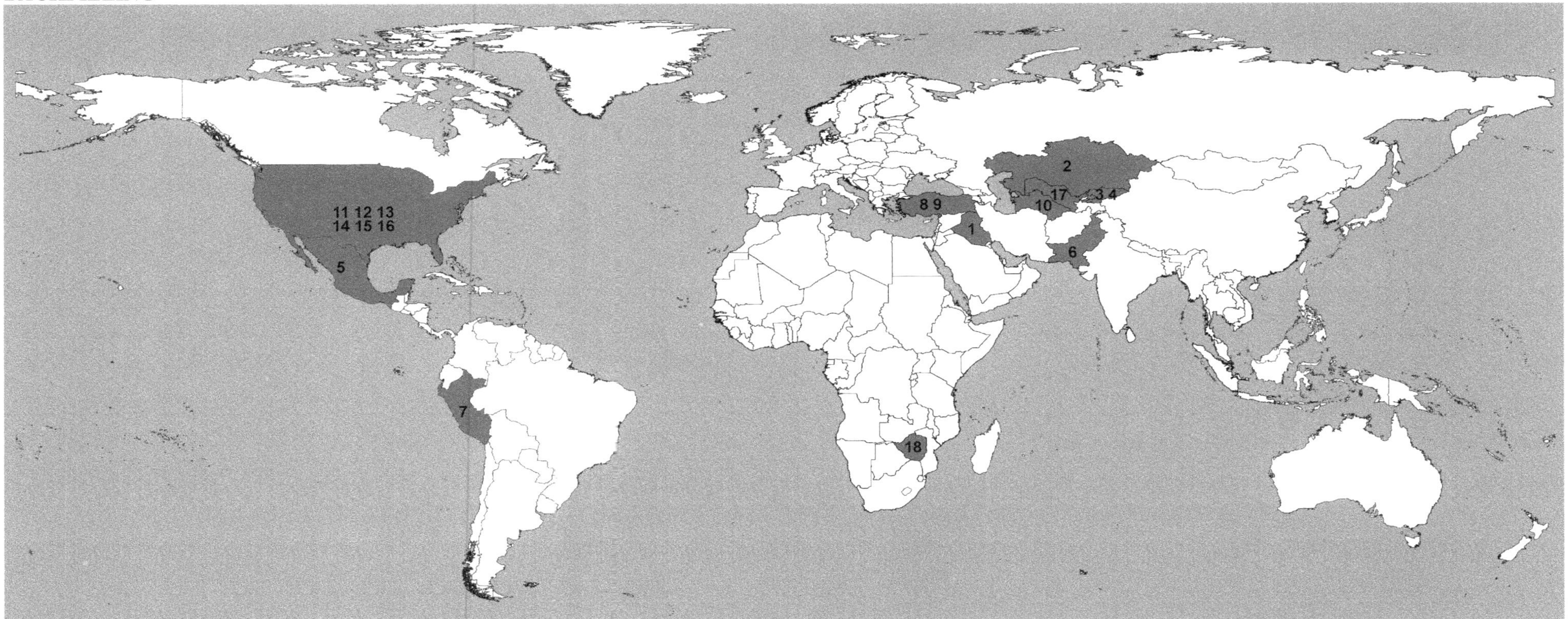

MAP ID	COUNTRY	SITE	MATERIALS	REFERENCE
1	Iraq	Tell Umar	Earth?	Chiari 1990a, 1990b
2	Kazakhstan	Otrar Tobe	Geotextile; earth	Fodde 2003
3	Kygystan	Ak Beshim	Geotextile; typar 32	Hurd 2003; Imankuluv and Tentieva 2003.
4	Kyrgystan	Navekat	Geotextile	Hurd 2003; Imankuluv and Tentieva 2003.
5	Mexico	Las Cuarenta Casas	Earth	Carrera 1993
6	Pakistan	Moenjodaro	Earth; geotextile	Hughes 1996; Jansen 2003.
7	Peru	Tomaval Castle	Earth	Hoyle *et al* 1993
8	Turkey	Çatalhöyük	Nothing; sandbags; vermiculite/perlite protection	Site Dossier: TURK0002

			system (temporary backfilling) Nothing; sandbags; geotextile (long-term backfilling)	
9	Turkey	Gordion	Geotextile, sand bags	Goodman 2002
10	Turkmenistan	Merv	Geotextiles; earth; sandbags (temporary backfilling) geotextiles; earth (long-term backfilling)	Site Dossier: TURM0001
11	USA	Aztec	Earth; geogrids; drains; geotextiles	Site Dossier: USAM0001
12	USA	Chaco	Earth; geogrids; drains; geotextiles	Site Dossier: USAM0003
13	USA	Pecos	Earth	Site Dossier: USAM0021
14	USA	Fort Selden State Monument	Earth	Caperton 1987; Agnew 1990
15	USA	Bandelier National Monument	Earth	Site Dossier: USAM0037
16	USA	San Diego Royal Presidio	Geotextile; earth	Calarco 2000
17	Uzbekistan	Sappali-Tepe	Earth?	Reutova and Shirinov 2004
18	Zimbabwe	Great Zimbabwe. (Posselt house)	Earth	Matsikure 2000

Table 5. Documented sites and materials used for backfilling
n.b see those sites with a site dossier for bibliographic references.

CAPPING

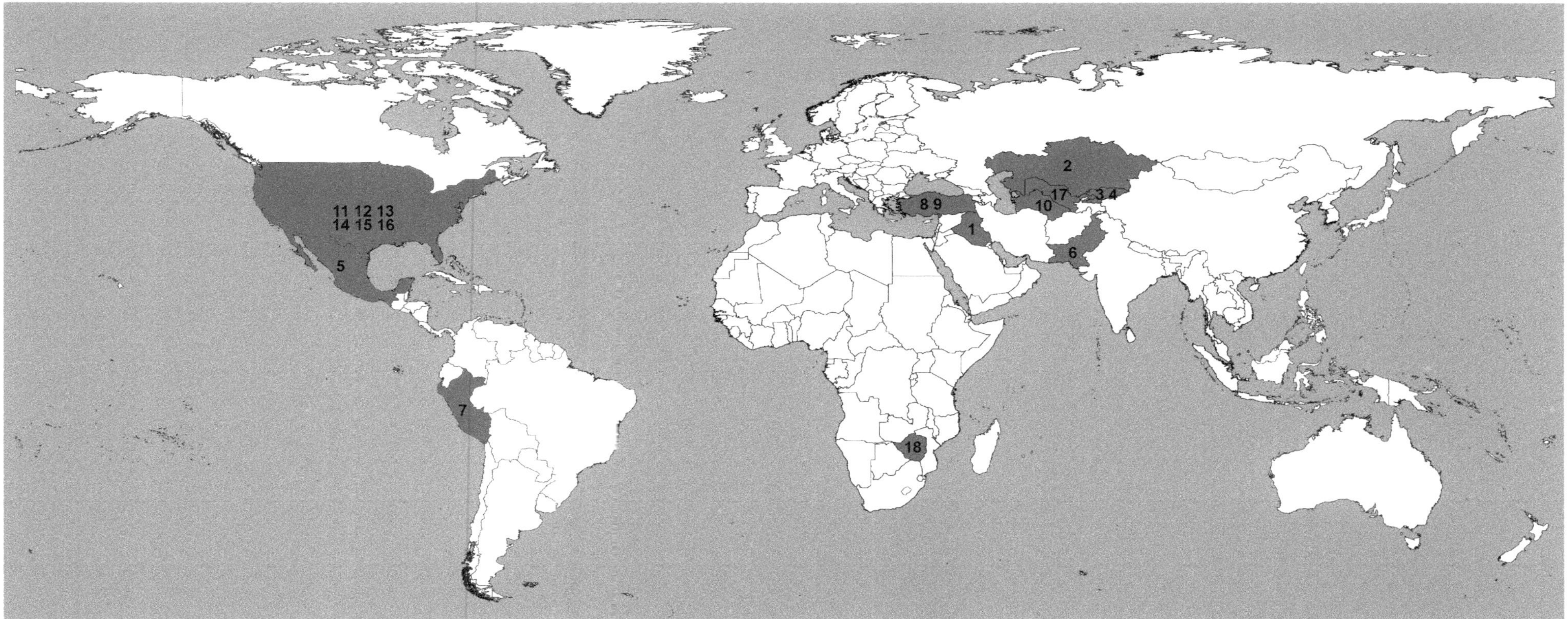

MAP ID	COUNTRY	SITE	MATERIALS	REFERENCE
1	Iraq	Tell ʻUmar (Seleucia)	Cement-amended modified mudbricks	Torroca, Chiari, and Gullini 1972
2	Iraq	ʻAqar Quf	Cement-amended modified mudbricks	Torroca, Chiari, and Gullini 1972
3	Iran	Persepolis	Mudbricks, mortars, plasters	Faccena 1976
4	Iran	Yazd	Cement and fired bricks.	Site Dossier: IRAN0007
5	Iran	Bam	Mudbricks, earth mortar, earth plaster	Site Dossier: IRAN0010
6	Iran	Rayen	Mudbricks, earth mortar, earth plaster	Site Dossier: IRAN0035
7	Iran	Shemsh	Mudbricks, earth mortar, earth plaster	Site Dossier: IRAN0036
8	Iran	Shahdad	Mudbricks, earth mortar, earth plaster	Site Dossier: IRAN0034
9	Oman	Khor Rori	Lime, sand and minimal cement mortar	Orazi 2000
10	Peru	Tomaval Castle	Mudbricks	Hoyle *et al* 1993
11	Turkmenistan	Merv	Cement and fired bricks.	Site Dossier: TURM0001
12	Turkmenistan	Nisa	Cement and fired bricks.	Site Dossier: TURM0002

13	Turkmenistan	Gonur	Cement; fired bricks; mudbricks; earth mortar; earth plaster.	Site Dossier: TURM0015
14	USA	Fort Selden	Pencapsula; acrylic polymer; amended capping	Selwitz 1995; Caperton 1987; Taylor 1990; Taylor 1987; Oliver 2000
15	USA	Fort Union	Cement-amended modified mudbricks	Hartzler and Oliver 2000
16	Uzbekistan	Khiva	Cement and fired bricks.	Site Dossier: UZBE0001
17	Uzbekistan	Shahrisabz	Cement and fired bricks.	Site Dossier: UZBE0003
18	Uzbekistan	Rabat-I-Malik	Cement and fired bricks.	Site Dossier: UZBE0011

Table 6. Documented sites and materials used for capping
n.b see those sites with a site dossier for bibliographic references.

ENCAPSULATION

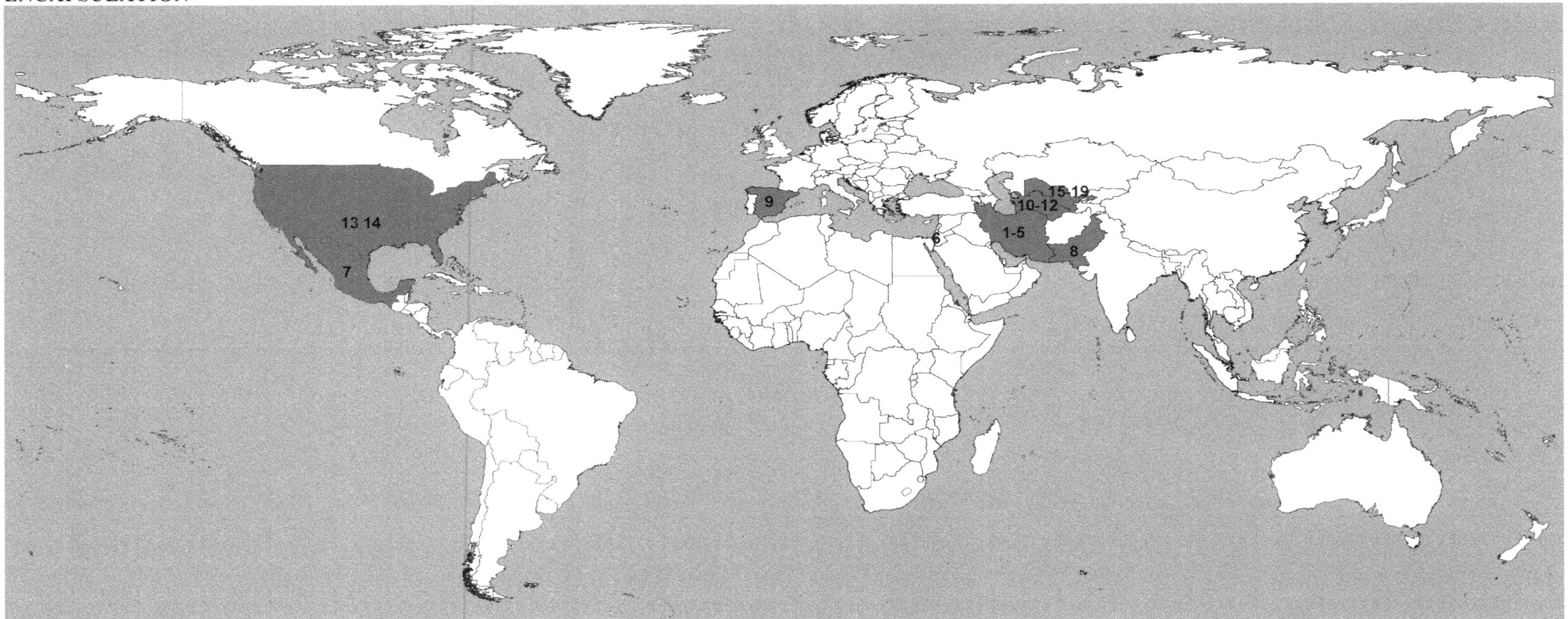

MAP ID	COUNTRY	SITE	MATERIALS	REFERENCE
1	Iran	Yazd	Cement, fired brick, glazed tiles, mudbrick, earth mortar, earth plaster	Site Dossier: IRAN0007
2	Iran	Bam	Mudbricks, earth mortar, earth plaster	Site Dossier: IRAN0010
3	Iran	Shahdad	Mudbricks, earth mortar, earth plaster	Site Dossier: IRAN0034
4	Iran	Rayen	Mudbricks, earth mortar, earth plaster	Site Dossier: IRAN0035
5	Iran	Shemsh	Mudbricks, earth mortar, earth plaster Cement, fired brick, breeze block	Site Dossier: IRAN0036
6	Israel	Tell Qasile	Mud plasters; amended mud plasters	Mazar 1999
7	Mexico	Paquime, Casa Grandes	Earthen materials	Brown *et al* 1990
8	Pakistan	Moenjodaro	Mud plasters; mudbricks	Hughes 1996; Jansen 2003.
9	Spain	Grenada	Cement (old conservation work)	Roca *et al* 1993
10	Turkmenistan	Merv	Cement, fired brick, mudbrick	Site Dossier: TURM0001

11	Turkmenistan	Nisa	Cement, fired brick, mudbrick	Site Dossier: TURM0002
12	Turkmenistan	Gonur	Cement; fired bricks; mudbricks; earth mortar; earth plaster; plastic sheeting; geotextile	Site Dossier: TURM0015
13	USA	Fort Selden	Amended materials	Selwitz 1995; Oliver 2000
14	USA	Pecos	New modified mudbricks	Site Dossier: USAM0021
15	Uzbekistan	Khiva	Cement, fired brick, glazed tiles.	Site Dossier: UZBE0001
16	Uzbekistan	Bukhara	Cement, fired brick, glazed tiles, paksha	Site Dossier: UZBE0002
17	Uzbekistan	Shahrisabz	Cement, fired brick, glazed tiles.	Site Dossier: UZBE0003
18	Uzbekistan	Samarkand	Cement, fired brick, glazed tiles,	Site Dossier: UZBE0004
19	Uzbekistan	Rabat-I-Malik	Cement, fired brick	Site Dossier: UZBE0011

Table 7. Documented sites and materials used for encapsulation.
n.b see those sites with a site dossier for bibliographic references.

CONSOLIDATION

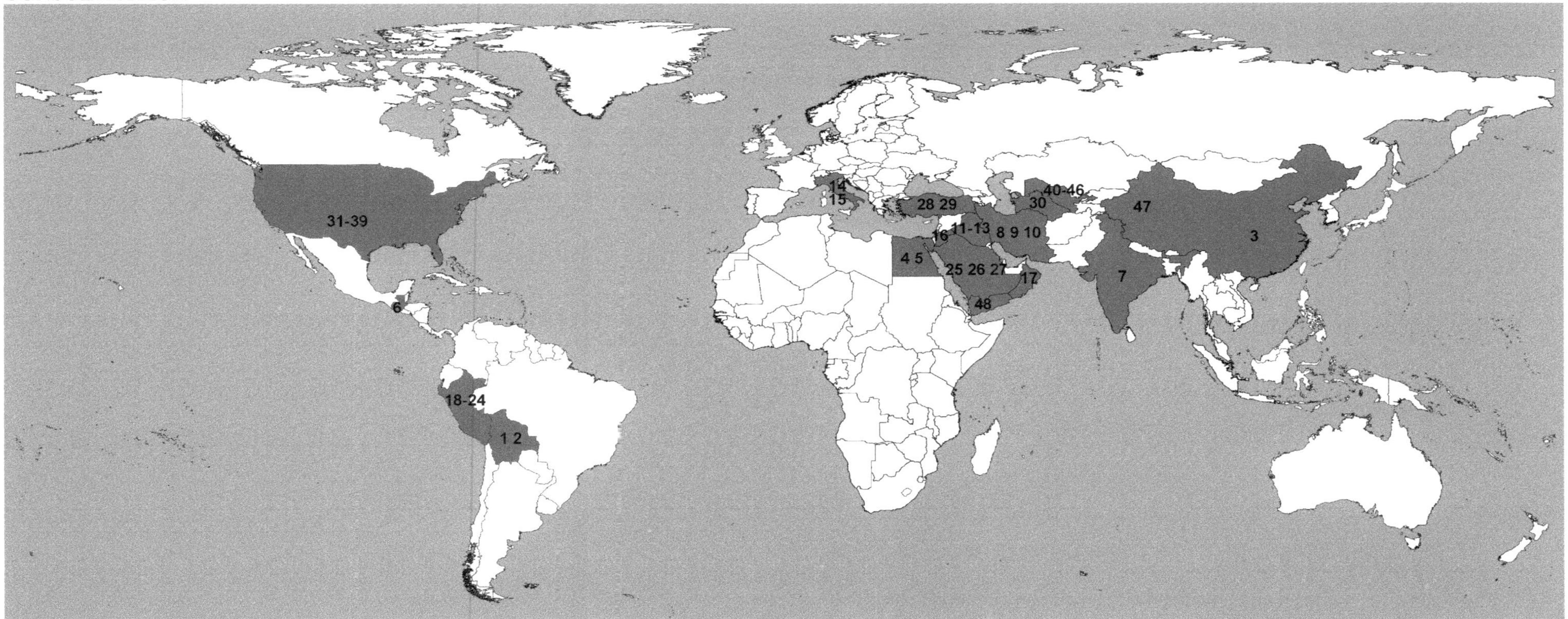

MAP ID	COUNTRY	SITE	MATERIALS	REFERENCE
1	Bolivia	Church at Carabuco	Acryloid B72; toulene	Rua *et al* 1993
2	Bolivia	Curahuara de Carangas	Polyvinyl alcohol, acryloid B-72	Rua and Rajer 1990
3	China	Dadiwan	Potassium silicate	Zuixiong 1990
4	Egypt	Abu-Sir	Paraloid B72; KP-LAK 709; wacker H; silgel JHM 2	Helmi 1990; Sramek and Losos 1990
5	Egypt	Thebes-West	Silicic acid ester (funcosil SAE 300); elasticised variant (Funcosil SAE 300E); silica-nanosol Sebosil S; Paraloid B72	Richter 2004
6	Guatemala	Chimaltenango	Meth-acrylic polymers; E-330 emulsion; 3% A-21, a combination treatment of 3% A-21 with a second treatment of 6% A-21), a combination 3% A-21 with E-330 as a plaster	Butterbaugh and Piggot 1980; Hartzler 1996
7	India	Basgo	Bitumen; kerosene; apricots	Gupta 2003

8	Iran	Tepe Nush-I Jan	Epikote; lacquer	Lewis 1980
9	Iran	Hasanlu Tepe	Meth-acrylic polymers;E-330 emulsion, 3% A-21, a combination treatment of 3% A-21 with a second treatment of 6% A-21, a combination 3% A-21 with E-330 as a plaster acrylic polymers; acryloid A-21; B-67	Piggot and Butterbaugh 1978; Butterbaugh and Piggot 1980
10	Iran	Esfahan	Sulfonated melamin formaldehyde; polystyrene foam	Langroudi 2003
11	Iraq	Uruk	Sodium silicate, calcium chloride, polyurethane resin, stabilised bricks.	Carter and Pagliero 1966
12	Iraq	Samarra	Ethyl silicate; synthetic polymers	Bruno *et al* 1968
13	Iraq	Tell Umar	Ethyl silicate; polyurethane resin; cement; silester ZNS; TEOS; wacker strengtheners OH	Torraca *et al* 1972; Chiari 1990a; Chiari 1990b
14	Italy	Feltre	Wacker strengthener OH; monsanto silester ZNS; ethyl silicate.	Chiari *et al* 1993
15	Italy	Crypta Baalbi	Earthen materials; hydraulic lime	Nardi 1987a; Nardi 1987b
16	Jordan	Teleilat Ghassul	Ethyl silicate	Schwartzbaum *et al* 1980
17	Oman	Khor Rori	Cement; lime; bitumen and chemical stabilisers (sulphonate petroleum products)	Orazi 2000
18	Peru	Tomaval Castle	0.05% vegetable binder (cacti mucilage: opuntia ficus indica	Hoyle *et al* 1993
19	Peru	Huaca Garagay	Acrylic emulsion, ethyl silicate, paraloid.	Chiari 1980
20	Peru	Cardal Lur	Tetra ethyl ortho silicate (TEOS); ethyl silicate; paraloid B72; wooden supports; Japanese rice paper	Chiari *et al* 2000
21	Peru	Chan Chan	Acrylic emulsion, organic agglutinates; ethyl silicate, paraloid.	Chiari 1980
22	Peru	Huaca del Dragon, Trujillo	Ethyl silicate (lab testing sample)	Chiari 1987
23	Peru	Nazca	Cola sinteca; imlar CPC	Skibinski 1990
24	Peru	Tulor	Ethyl silicate; wacker strengthener OH	Muoz and Bahamondez 1990
25	Saudi Arabia	Masmak, Riyadh	Ethyl silicate; concrete	Albini 1980
26	Saudi Arabia	al-'Udhaibat	Concrete; durspan 4TC; ethyl silicate	Othman 2003
27	Saudi Arabia	Royal Palace of Muraba	Concrete; durspan 4TC; ethyl silicate	Othman 2003
28	Turkey	Çatalhöyük	PVA	Site Dossier: TURK0002
29	Turkey	Gordion	PVA; acrylic resins; acrysol WS-24, acryloid B-72	Koob *et al* 1990
30	Turkmenistan	Nisa	Isocyanate monomers, ethyl silicate, paraloid B72; paraloid, polyfilla interior; dental plaster	Site Dossier: TURM0002
31	USA	Aztec	Shellac coating; cellulose nitrate; aAcrylic modified earths; isopropyl alcohol	Site Dossier: USAM0001
32	USA	Fort Selden	Polyurethane; polymers, polyisocyanates; surface coats	Coffman *et al* 1990; Agnew *et al* 1987; Selwitz 1995; Oliver 2000

33	USA	Chaco	Acrylic modified earths; acrylic emulsions; rhoplex E-863, E-330, E-826	Site Dossier: USAM0003
34	USA	Casa Grande National Park	Acrylic emulsions; grouts; lime; binder; modifier; acrylic el rey superior 200 emulsion; meth-acrylic polymers; E-330 emulsion, 3% A-21, a combination treatment of 3% A-21 with a second treatment of 6% A-21, a combination 3% A-21 with E-330 as a plaster	Cancino and Matero 2003
35	USA	Tumacacori	Acrylic emulsions; rhoplex E-863, E-330, E-826	Crosby 1980
36	USA	Fort Union National Monument	Acrylic emulsions; water repellents	Hartzler and Oliver 2000
37	USA	Bents Old Fort	Acrylic emulsions; synthetic latex soil slurry; airflex 510, UCAR 365, acryloid F-10.	Ferm 1990
38	USA	Pio Pico Mansion Adobe	Lime; fly ash; portland cement; modified earth	Roselund 1990
39	USA	Pecos	Acrylic modified earth; acrylic emulsions; acrylic modified earthen mortars (rhoplex E-330)	Site Dossier: USAM0021
40	Uzbekistan	Samarkand	Isocyanate monomers; di-isocyanates;	Site Dossier: UZBE0004
41	Uzbekistan	Fayas Tepe, Termez	Polymer, monomer, ethyl silicate, isocyanate,	Abdurazakov 1986
42	Uzbekistan	Akh-Tepe, Tashkent	Di-isocyanates	Abdurazakov 1986
43	Uzbekistan	Sapilii Tepe	Di-isocyanates	Abdurazakov 1986
44	Uzbekistan	Kara Tepe	Di-isocyanates	Abdurazakov 1986
45	Uzbekistan	Dzarkutan	Di-isocyanates	Abdurazakov 1986
46	Uzbekistan	Kanka	Di-isocyanates	Abdurazakov 1986
47	Xinjaing	Kezier Grottoes	Potassium silicate; magnesium fluorosilicate; silanes; methyltriethoxysilane;	Kezhong 1990
48	Yemen	Sanaa	Hydraulic stabilizer	Olivier et al 1990

Table 8. Documented sites and materials used for consolidation.
n.b see those sites with a site dossier for bibliographic references.

DRAINAGE

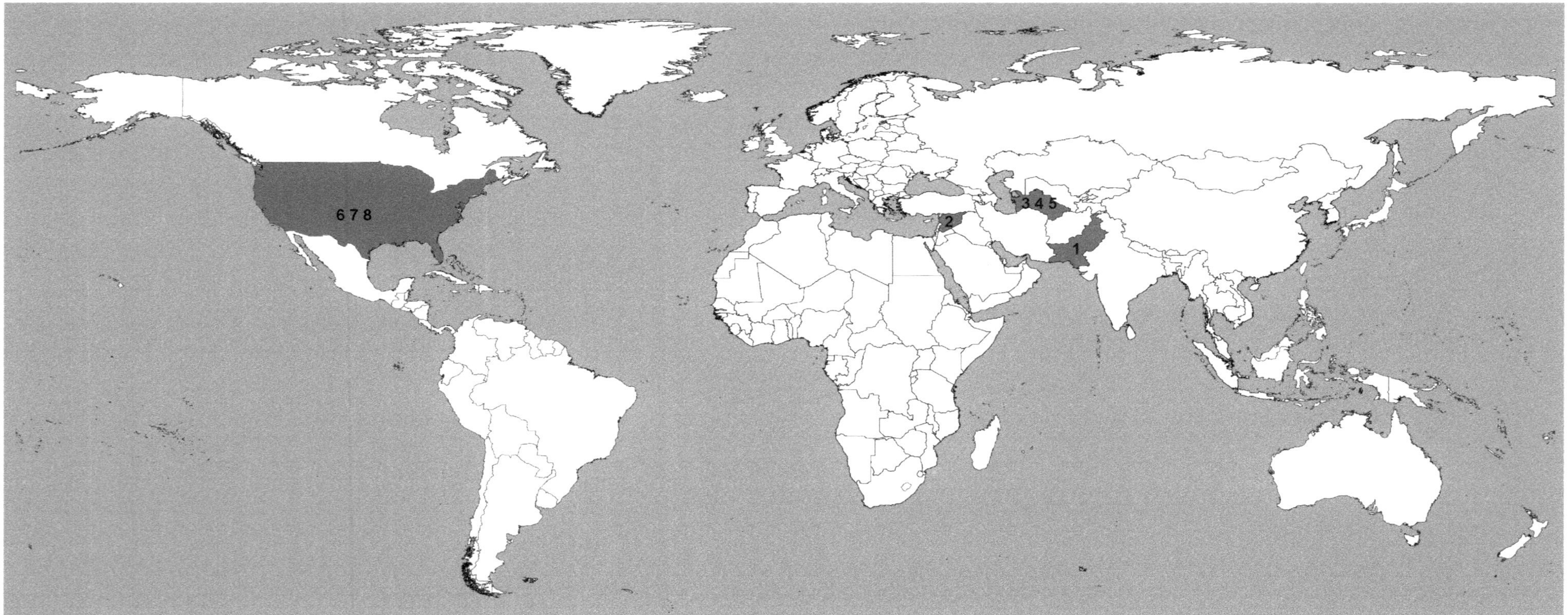

MAP ID	COUNTRY	SITE	MATERIALS	REFERENCE
1	Pakistan	Moenjodaro	Damp-proof coursing	Hughes 1996; Jansen 2003.
2	Syria	Mari	Surface drains	Bendakir 1993; *pers comm.* Mahmoud Bendakir
3	Turkmenistan	Merv	Aggregate; geotextile; mudbrick; drainage slopes (with backfilling)	Site Dossier: TURM0001
4	Turkmenistan	Nisa	Aggregate; surface drains	Site Dossier: TURM0002
5	Turkmenistan	Gonur	Aggregate; geotextile	Site Dossier: TURM0015
6	USA	Fort Selden	Drainage slopes	Caperton 1987, 1990, 1993; Agnew 1990; Oliver 2000.
7	USA	Aztec	Surface drains	Site Dossier: USAM0001
8	USA	Chaco	Surface drains	Site Dossier: USAM0003

Table 9. Documented sites and materials used for drainage. n.b see those site with a site dossier for bibliographic references.

UNDERCUT REPAIR

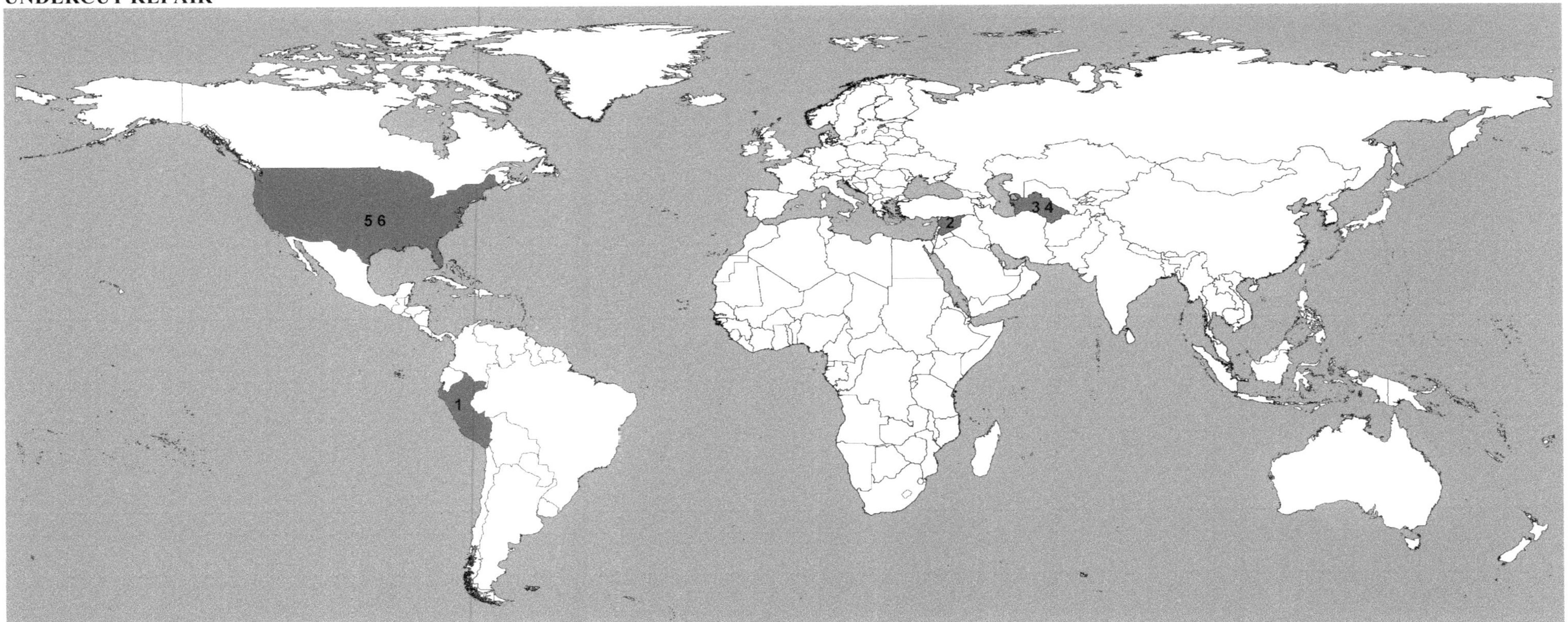

MAP ID	COUNTRY	SITE	MATERIALS	REFERENCE
1	Peru	Tomaval Castle	Earth	Hoyle *et al* 1993
2	Syria	Mari	Geotextile, earth infill	Bendakir 1993; pers comm. Mahmoud Bendakir
3	Turkmenistan	Merv	Geotextile; fired brick; cement mortars; mudbrick; earthen mortars; placed earth; rammed earth	Site Dossier: TURM0001
4	Turkmenistan	Gonur	Geotextile; fired brick; cement mortars; mudbrick; earthen mortars	Site Dossier: TURM0015
5	USA	Fort Selden	Amended mudbricks	Caperton 1990, 1993; Oliver 2000
6	USA	Casa Grande National Park	Cement, amended earth	Matero 1999; Matero et al 2000

Table 10. Documented sites and materials used for undercut repairs.
n.b see those site with a site dossier for bibliographic references.

MAINTENANCE

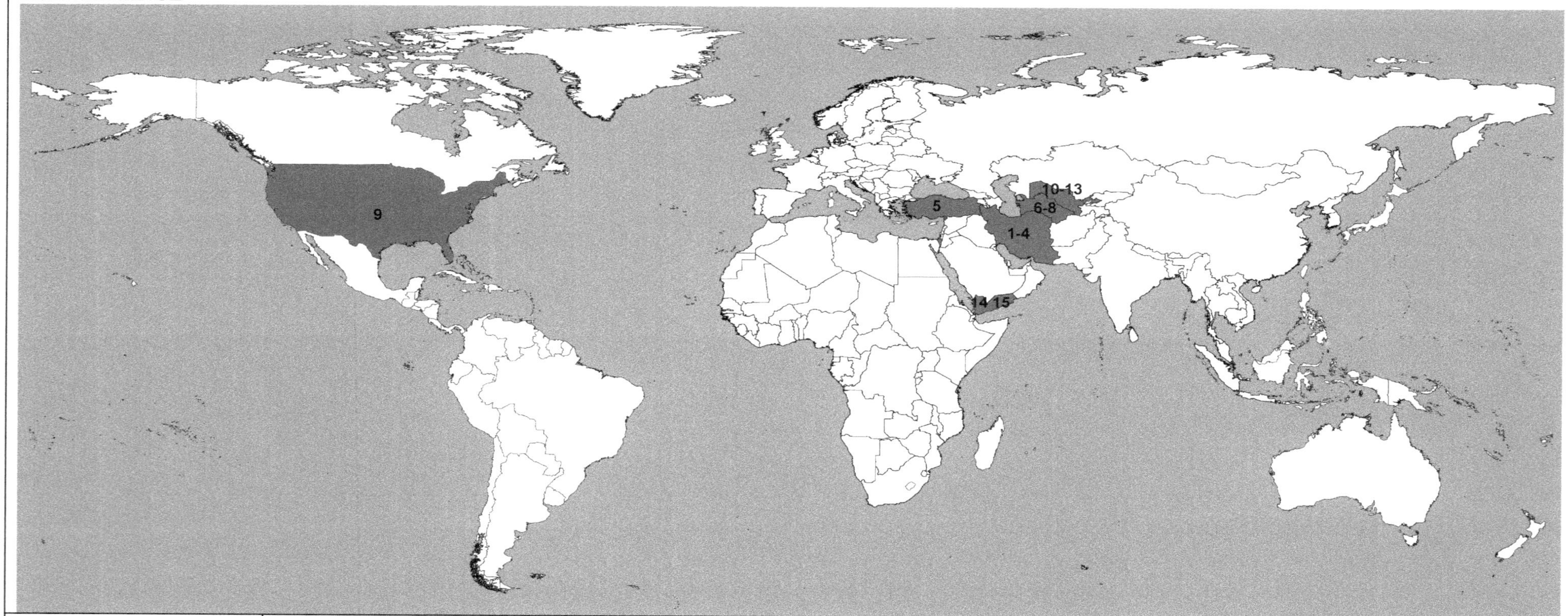

MAP ID	COUNTRY	SITE	MATERIALS	REFERENCE
1	Iran	Yazd	Cement, earth plaster, earth mortar	Site Dossier: IRAN0007
2	Iran	Shahdad	Earth plaster, earth mortar	Site Dossier: IRAN0034
3	Iran	Rayen	Earth plaster, earth mortar	Site Dossier: IRAN0035
4	Iran	Abianeh	Earth plaster, earth mortar	Dekhordi 2004
5	Turkey	Konya	Cement render	Site Dossier: TURK0006
6	Turkmenistan	Nisa	Earth plaster, earth mortar	Site Dossier: TURM0002
7	Turkmenistan	Gonur	Earth plaster, earth mortar	Site Dossier: TURM0015
8	Turkmenistan	Taklahtan Baba	Earth plaster	Site Dossier: TURM0018
9	USA	Aztec Ruins	Cements, bitumen, lime mortars	Site Dossier: USAM0001
10	Uzbekistan	Khiva	Earthen plasters, earthen render, cement	Site Dossier: UZBE0001
11	Uzbekistan	Bukhara	Cement, earthen plaster; earthen	Site Dossier: UZBE0002

			mortar	
12	Uzbekistan	Shahrisabz	Cement, earthen plaster; earthen mortar	Site Dossier: UZBE0003
13	Uzbekistan	Samarkand	Cement	Site Dossier: UZBE0004
14	Yemen	Shibam	Lime plaster; earth plaster	Damluji 1992
15	Yemen	Tarim	Lime plaster; earth plaster	Jerome 2000; Jerome et al 2003

Table 11. Docummented sites and materials used for maintainance.
n.b see those site with a site dossier for bibliographic references.

RECONSTRUCTION

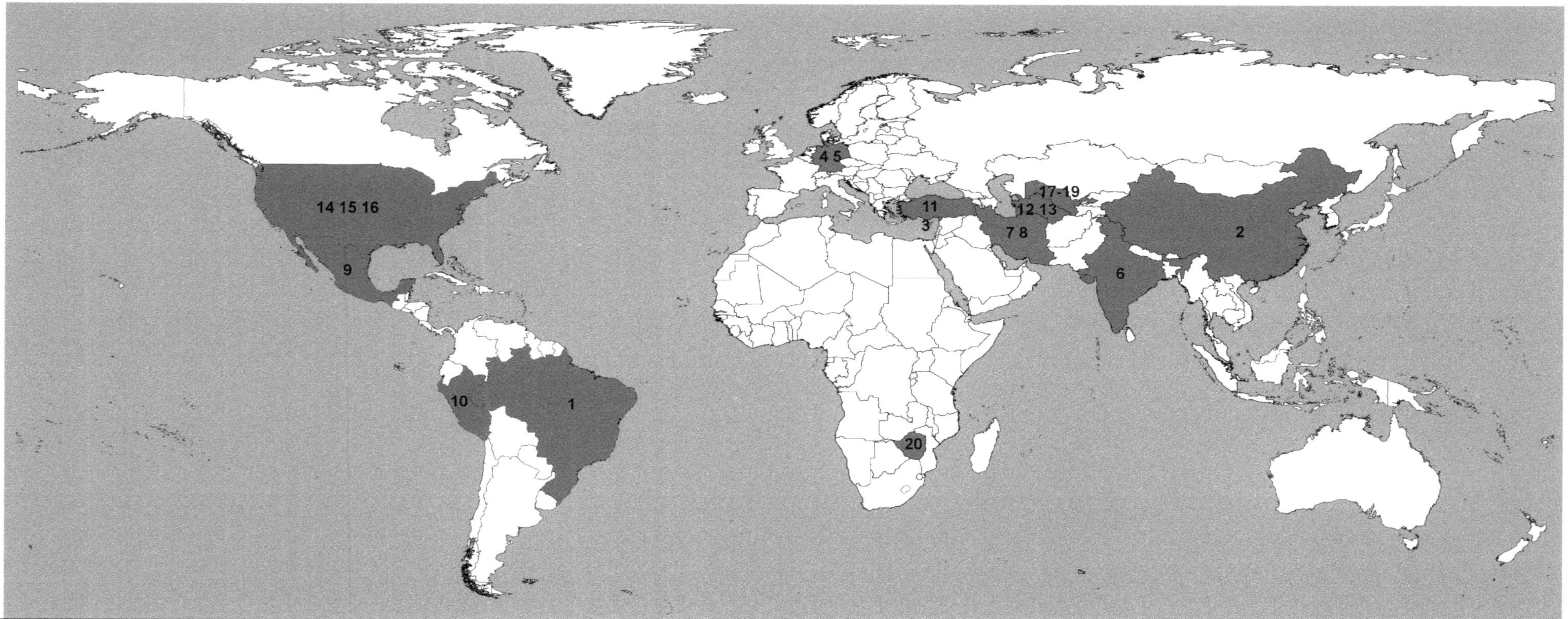

MAP ID	COUNTRY	SITE	MATERIALS	REFERENCE
1	Brazil	Novo Hamburgo	Unspecified	Soliani et al 1993
2	China	Mangshan	Unspecified	Jiyao and Weitung 1990
3	Cyprus	Lemba	Ex situ reconstruction of roundhouses	Thomas 1999
4	Germany	Chapel of Reconstruction	Rammed earth (reusing materials).	Site Dossier: GERM0001
5	Germany	Heuneburg	Earth blocks	Seefried 2004
6	India	Kusinara	Concrete	Sengupta 1984
7	Iran	Yazd	Earthen materials	Site Dossier: IRAN0007
8	Iran	Rayen	Earthen materials	Site Dossier: IRAN0035
9	Mexico	Las Cuarenta Casas	Unspecified	Carrera 1993
10	Peru	Tulor	Earthen materials; modified earthen materials	Muoz and Bahamondez 1990
11	Turkey	Çatalhöyük	Ex situ reconstruction of neolithic house.	Site Dossier: TURK0002

12	Turkmenistan	Merv	Fired bricks, mudbrick, earthen plasters	Site Dossier: TURM0001
13	Turkmenistan	Nisa	Fired bricks, mudbricks, white wash	Site Dossier: TURM0002
14	USA	Aztec Ruins	*In situ* reconstruction of kiva	Site Dossier: USAM0001
15	USA	Pecos	*In situ* reconstruction of kiva	Site Dossier: USAM0021
16	USA	Bandelier	*In situ* reconstruction of kiva etc	Site Dossier: USAM0037
17	Uzbekistan	Khiva	*In situ* stretch of city wall.	Site Dossier: UZBE0001
18	Uzbekistan	Bukhara	*In situ* stretch of city wall	Site Dossier: UZBE0002
19	Uzbekistan	Shahrisabz	*In situ* stretch of city wall.	Site Dossier: UZBE0003
20	Zimbabwe	Great Zimbabwe	*In situ* reconstruction (on top of backfilled zone)	Matsikure 2000

Table 12. Documented sites and materials used for reconstruction.
n.b see those sites with a site dossier for bibliographic references.

RESTORATION

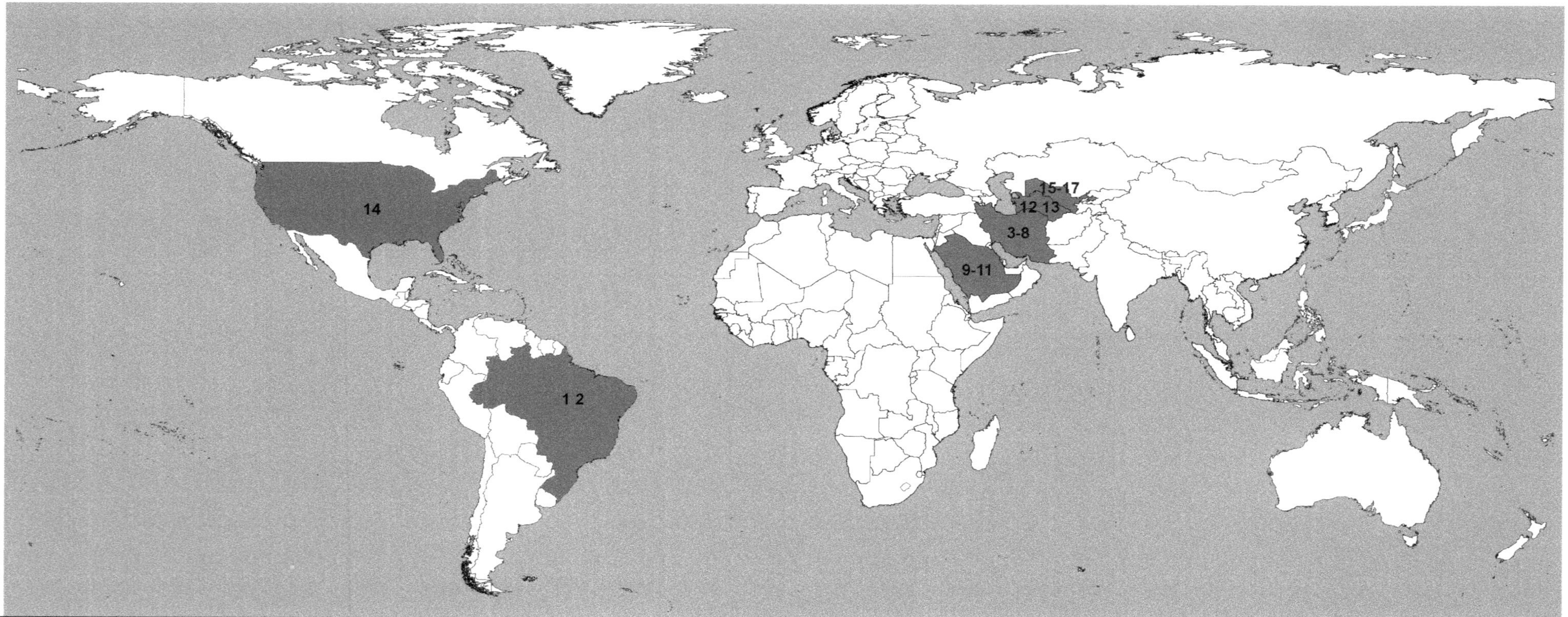

MAP ID	COUNTRY	SITE	MATERIALS	REFERENCE
1	Brazil	Basilica of Our Lady of Pillar	Unspecified	Lima and Puccioni, 1990
2	Brazil	Church of Nossa Senhora do Rosario	Unspecified (cement in the past)	Pecararo 1993
3	Iran	Kashan	Adaptive re-use, materials unspecified	Azghandi 2003
4	Iran	Yazd	Adaptive re-use, various: see dossier	Site Dossier: IRAN0007
5	Iran	Bam	(As above)	Site Dossier: IRAN0010
6	Iran	Shahdad	(As above)	Site Dossier: IRAN0034
7	Iran	Rayen	(As above)	Site Dossier: IRAN0035
8	Iran	Shemsh	(As above)	Site Dossier: IRAN0036
9	Saudi Arabia	Dir'iyah	Earth materials	Emrick and Meinhardt 1990
10	Saudi Arabia	Masmak, Riyadh	Earth materials but ethyl silicate of surface	Albini 1980

11	Saudi Arabia	al-'Udhaibat	Concrete; durspan 4TC; ethyl silicate;	Othman 2003
12	Turkmenistan	Merv	Various: see dossier	Site Dossier: TURM0001
13	Turkmenistan	Anau	Various: see dossier	Site Dossier: TURM0017
14	USA	Acoma	Various: earthen materials, cement	McHenry 1990
15	Uzbekistan	Khiva	Various: see dossier	Site Dossier: UZBE0001
16	Uzbekistan	Bukhara	Various: see dossier	Site Dossier: UZBE0002
17	Uzbekistan	Samarkand	Various: see dossier	Site Dossier: UZBE0004

Table 13. Documented sites and materials used for restoration.
n.b see those sites with a site dossier for bibliographic references.

REMOVAL/RELOCATION

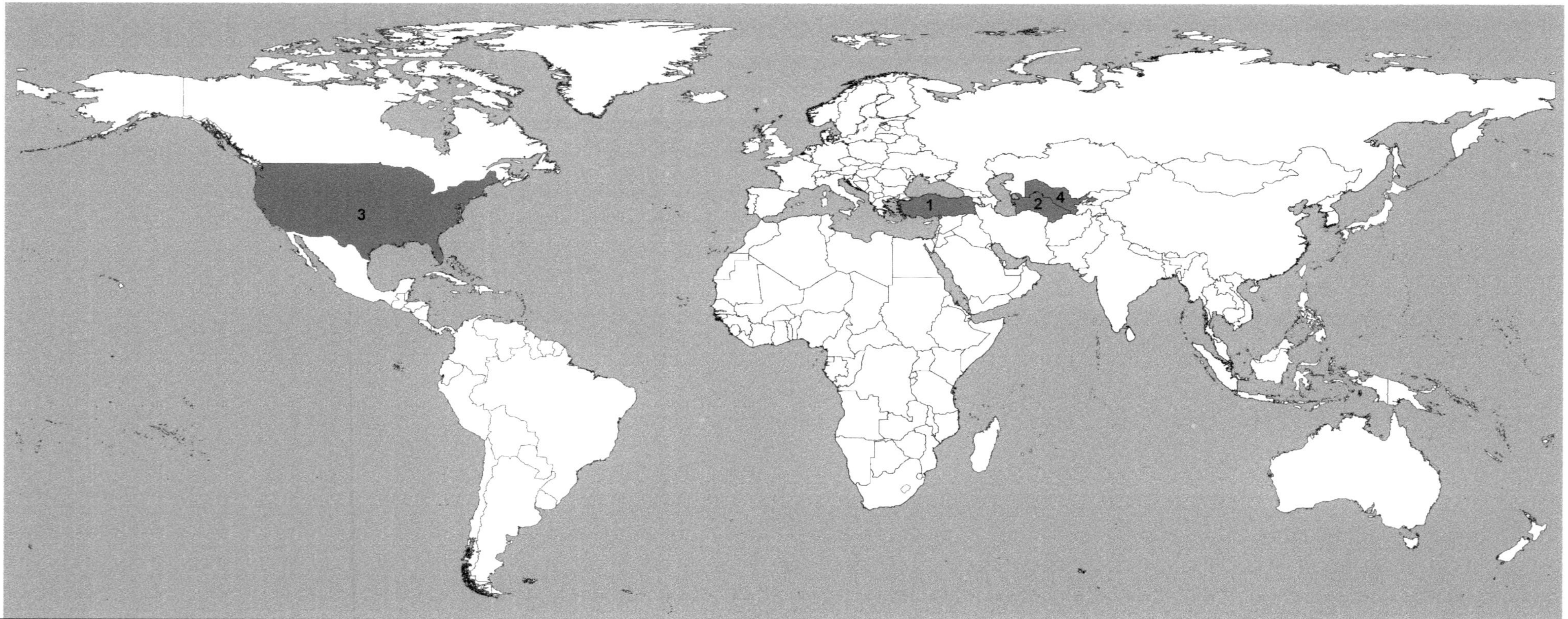

MAP ID	COUNTRY	SITE	MATERIALS	REFERENCE
1	Turkey	Çatalhöyük	Consolidation and display in museum	Site Dossier: TURK0002
2	Turkmenistan	Nisa	Ethyl silicate, paraloid B72; 3% paraloid in trichorethane; polyfilla interior; dental plaster	Site Dossier: TURM0002
3	USA	Escalante Ruin	Testing consolidants and detachment methodologies	Silver 1990
4	Uzbekistan	Samarkand	Consolidation and display in museum	Site Dossier: UZBE0004

Table 14. Documented sites and materials used for removal/relocation.

SHELTERING

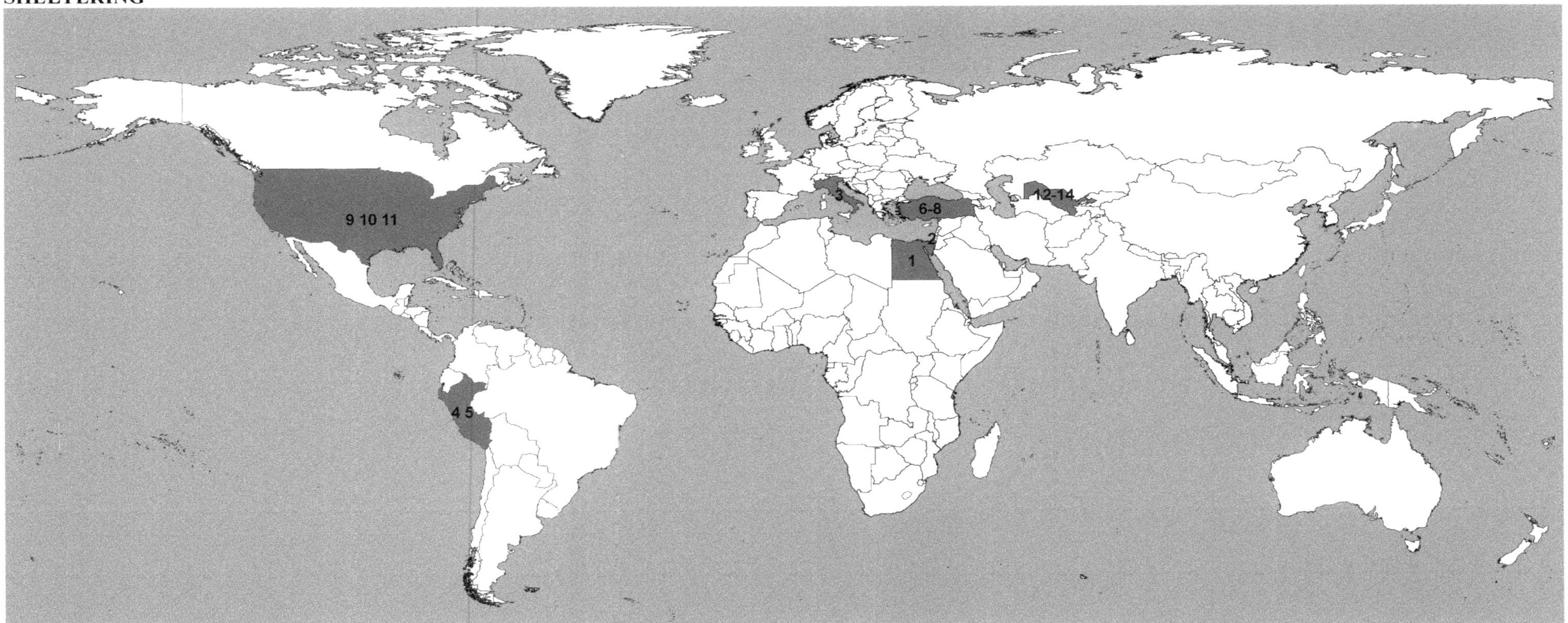

MAP ID	COUNTRY	SITE	MATERIALS	REFERENCE
1	Egypt	Abu-Sir	Unspecified (used with consolidation of fabric)	Sramek and Losos 1990
2	Israel	Tell Qasile	Various for 'roof' erection	Mazar 1999
3	Italy	Crypta Baalbi	Unspecified (used with consolidation of fabric)	Nardi 1987a, 1987b
4	Peru	Chan Chan	Semi-permanent natural materials	Chiari 1980
5	Peru	Huaca Garagay	Semi-permanent natural materials	Chiari 1980
6	Turkey	Kaman-Kalehoyuk	Timber	Carroll 1998
7	Turkey	Çatalhöyük	Plasticised tarpaulin and metal; lightweight metal space frame; canvas/tarpaulin and timber supports;	Site Dossier: TURK0006
8	Turkey	Konya	Concrete dome	Site Dossier: TURK0006
9	USA	Aztec Ruins	Cement/ tar paper/ felt	Site Dossier: USAM0001
10	USA	Casa Grande National Park	Redwood	Matero 1999; Matero et al 2000; Rael 2004
11	USA	Bandalier	Perspex cover	Site Dossier: USAM0037

12	Uzbekistan	Shahrisabz	Corrugated plastic	Site Dossier: UZBE0003
13	Uzbekistan	Samarkand	Corrugated plastic	Site Dossier: UZBE0004
14	Uzbekistan	Sapilii Tepe	Unspecified	Reutova and Shirinov 2004

Table 15. Documented sites and materials used for sheltering.
n.b see those sites with a site dossier for bibliographic references.

Bibliography

Abdurazakov, A. 1986. Conservation experience of clay construction in Uzbekistan. *Preventative Measures during excavation and site protection, Conference, Ghent 6-8 November 1985.* 81-90. ICCROM.

Agnew, N. 1990. The Getty adobe research project at Fort Selden I. Experimental design for a test wall project. In *6th International Conference on the Conservation of Earthen Architecture, Adobe 90 Preprints, Las Cruces, New Mexico, USA, October 14-19, 1990.* 243-249. Los Angeles: The Getty Conservation Institute.

Agnew, N, Druzik, J, Caperton, T and Taylor, M. 1987. Adobe: the earliest composite material. In *ICOM committee for conservation, 8th triennial meeting, Sydney, Australia, 6th to 11th September, 1987. Preprints.* 439-446. Los Angeles: The Getty Conservation Institute.

Agnew, N and Demas, M. 2004. *Principles for the Conservation of Heritage Sites in China, English-language Translation, with Chinese Text, of the Document Issued by China ICOMOS.* Los Angeles: The Getty Conservation Institute.

Albini, M. 1980. Mudbrick Buildings. The Masmak Fortress in Riyadh – Saudi Arabia general criteria of restoration. *Third International Symposium on Mudbrick (adobe) Preservation, 29th September to 4th October 1980, Ankara, Turkey.* 119-138. ICOM/ICOMOS.

Alva, A, Torraca, G and Uginet, M. 1980. Adobe Bibliography. *Third International Symposium on Mudbrick (adobe) Preservation, 29th September to 4th October 1980, Ankara, Turkey.* 241-244. ICOM/ICOMOS.

Alva, A, Doat, P, Guillaud, H, Houben, H, Joffrey, T, Odul, P, Teutonico, J and Trappeniers, M. 1990. CRATerre-EAG, ICCROM long-term plan for the preservation of the earthen architectural heritage: The Gaia Project. *6th International Conference on the Conservation of Earthen Architecture, Las Cruces, New Mexico, USA. October 14-19, 1990. Preprints.* 461-468. Los Angeles: The Getty Conservation Institute.

Architecture for Humanity. 2006. *Design Like You Give Damn.* London: Thames and Hudson.

Argumendo, R. 1981. The Restoration of mud-brick structures and historical monuments of the Andean region of Peru. In *Appropriate Technologies in the Conservation of Cultural Property.* 83-113.Technical Handbooks for Museums and Monuments. UNESCO.

Ashurst, J and Ashurst, N. 1988. *Practical Building Conservation Volume 2. Brick, Terracotta and Earth.* English Heritage Technical Handbook. London: English Heritage.

Aslan, Z. 1997. Protective Structures for conservation and presentation of archaeological sites. *Journal of Conservation and Museum Studies.* No. 3. http://www.jcms.ucl.ac.uk/issue3/aslan.html

Avrami, E, Mason, R and de la Torre, M. 2000. *Values and Heritage Conservation Research Report.* Los Angeles: The Getty Conservation Institute.

Azghandi, A. 2003. Conservation – Restoration & Revitalization in Adobe Architecture A look at Kashan Ameriha Complex. In *Preprint of Papers of the 9th International Conference on the Study and Conservation of Earthen Architecture, Terra 2003, Yazd, Iran. 29th November – 2nd December 2003.* 48-59. Tehran: Iranian Cultural Heritage Organisation.

Azzout, M. 1999. The Soviet Interpretation and Preservation of the Ancient Heritage of Uzbekistan: The Example of Bukhara. In Petruccioli, A (ed). *Bukhara: The Myth and the Arch*itecture. 161-173. Cambridge, Massachusetts: The Aga Khan Program for Islamic Architecture.

Balderrama, A. 2001. The Conservation of Earthen Architecture. *The Getty Conservation Institute Newsletter.* Volume 16, No. 1. 4-10.

Balderrama, A and Chiari, G.1995 (1984). Protection and Conservation of Excavated Structures of Mudbrick. In Stanley Price, N (ed). *Conservation on archaeological excavations, with particular reference to the Mediterranean area.* 101-112. Rome: ICCROM. (2nd Edition).

Balderrama, A and Albertini, C. 2004. ICCROM's activities on the Study and the Conservation of Earthen Architecture. In *Lehm 2004. 4. Internationale Fachtagung für Lehmbau 29-30 October 2004. Leipzig, Germany.* 124-135. Weimar Germany: Dachverband Lehm e.V.

Baradan, B. 1990. A New Restoration Material for Adobe Structures. *6th International Conference on the Conservation of Earthen Architecture, Las Cruces, New Mexico, USA. October 14-19, 1990 Preprints.* 149-152. Los Angeles: The Getty Conservation Institute.

Barker, P. 1977. *Techniques of archaeological excavation.* London: Batsford.

Bastien, J. 1998. *The Kiss of Death – Chagas' Disease in the Americas.* Salt Lake City: The University of Utah Press.

Batsford, H and Fry, C. 1938. *The English Cottage.* London: Batsford.

Bazara, A. 2004. The conservation and renovation of earthen buildings in Wadi Haadramaut. Yemen. In Lehm 2004. *4. Internationale Fachtagung für Lehmbau 29.-30*

October 2004. Leipzig, Germany. 136-149. imar Germany: Dachverband Lehm e.V.

Beas, M. 1993. Consolidation of Traditional Plasters: A Laboratory Research. *7a conferência internacional sobre e estudo e conservação da arquitectura de terra, Silves, Portugal, 24-29 October 1993.* 410-416. Portugal: Direccao Geral dos Edificios e Monumentos Nacionais.

Beazley, E and Harverson, M. 1982. *Living with the desert. Working Buildings in the Iranian Plateau.* Warminster: Aris & Philips Ltd.

Bedaux, R. Diaby, B. Maas, P and Sidibe, S. 2000. The restoration of Jenné, Mali. African Aesthetics and Western Paradigms. *Terra 2000: 8th International Conference on the study and conservation of earthen architecture. Preprints. Torquay, United Kingdom, 11-13 May 2000.* 201-207. London: English Heritage.

Bedell, G. 2005. *The Handmade House.* London: Penguin Books.

Bendakir, M. 1993. Methodologie de recherché pour la preservation du site archeologique de Mari (Syrie). *7a conferência internacional sobre e estudo e conservação da arquitectura de terra, Silves, Portugal, 24-29 October 1993.* 317-323 Portugal: Direccao Geral dos Edificios e Monumentos Nacionais.

Best, H. 1642. *The farming and Memorandum Books of Henry Best of Elmswell.* D Woodward (ed). 1984. Records of Social and Economic History. Oxford: Oxford University Press for the British Academy.

Blackwell, C. 2001. *Tradition and Society in Turkmenistan.* Richmond: Curzon.

Blanchette, R. Held, B. Jurgens, J and Haight, J. 2004. Wood deterioration in Chacoan great houses of the southwestern United States. In *Conservation and Management of Archaeological Sites*, Vol. 6. 203-212.

Blondet, M and Garcia, G. 2003. Research Studies on Adobe Buildings and Earthquake. Preprints. *Moderner Lehmbau 2003.*139-147. Nachhaltiger Wohnungsbau – Zukunft Okologisches Bauen. Fraunhofer IRB Verlag.

Bourgeois, J and Pelos, C. 1996. *Spectacular Vernacular, The Adobe Tradition.* 2nd Edition. New York: Aperture.

Boyer, J. 1990. The House that Juan built: an archaeological perspective on adobe analysis. *6th International Conference on the Conservation of Earthen Architecture, Las Cruces, New Mexico, USA. October 14-19, 1990. Preprints.* 197-203. Los Angeles: Getty Conservation Institute.

Bradley, R. 2002. *The Past in Prehistoric Societies.* London: Routledge.

Branigan, T and Whitaker, B. 2003. *Dangerous buildings, lax rules: why Bam death toll was so high* http://www.guardian.co.uk/iran/story/0,12858,1112938,00.html

Braun, U. 2003. *Versöhnungskirche Kappelle der Versöhnung in Berlin.* (translated by D. Huyssen). Berlin: Evangelische Versöhnung Germinde.

Brooks, C. 1998. Introduction: Historicism and the Nineteenth Century, in V Brand (ed). *The Study of the Past in the Victorian Age,* 1-20. Oxford: Oxbow Books.

Brown, J and Bryant, R. 1999. *A Garden and Three Houses.* Woodbridge: Garden Art Club.

Brown, W and Clifton, J. 1978. Adobe 1 - the properties of adobe. *Studies in Conservation* 23. 139-146.

Brown, W. Robbins, C and Clifton, J. 1979. Adobe 2 - the properties of adobe. *Studies in Conservation* 24. 23-39.

Brown, R. Sandoval, B and Orea, H. 1990. The protection and conservation of the Adobe Structures at Paquime, Casa Grandes, Chihuahua, Mexico. *6th International Conference on the Conservation of Earthen Architecture, Las Cruces, New Mexico, USA. Preprints. October 14-19,* 1990. 204-208. Los Angeles: Getty Conservation Institute.

Brun, P and Annaev, A. 2002, contribution to, Williams, T and Kurbansakhatov, K *et al.* The Ancient Merv Project, Turkmenistan Preliminary Report on the 1st Season (2001). *Iran* 40 (2002). 1-28.

Brun, P and Annaev, A. 2003, contribution to, Williams, T and Kurbansakhatov, K *et al.* The Ancient Merv Project, Turkmenistan Preliminary Report on the 2nd Season (2002). *Iran* 41 (2003). 139-170.

Bruno, A and Chiari, G. 1968-69. Contributions to the study of the preservation of mud-brick structures. *Mesopotamia.* III-IV. 443-473.

Brunskill, R. 1981. *Traditional Buildings of Britain: an introduction to vernacular architecture.* London: Orion.

Burman, P. 1999. Philosophy and ethics for conserving earthen buildings. Paper presented at: Repair Methods for earthen buildings: The state of the art. An International Seminar at the King's Manor, York, Tuesday 18 May 1999.

Burman, P. 2003. A Question of Ethics. Article posted on http://www.buildingconservation.com/articles/ethics/ethics.htm (accessed November 2004).

Butterbaugh, D and Piggot, V. 1980. Masca Mudbrick/ Adobe conservation interim report. *Third International Symposium on Mudbrick (adobe) Preservation, 29th*

September to 4th October 1980, Ankara, Turkey. 19-28. ICOM/ICOMOS.

Byron, R. 1937. *The Road to Oxiana*. London: John Murray.

Byrne, D. 1991. Western Hegemony in Archaeological Heritage Management. *History and Anthropology, 5*, 269-276.

Calarco, D. 2000. San Diego Royal Presidio Conservation of an earthen archaeological site. *Terra 2000: 8th International Conference on the study and conservation of earthen architecture. Preprints. Torquay, United Kingdom, 11-13 May 2000*. 19-25. London: English Heritage.

Cameron, C and Tomka, S. *The Abandonment of Settlements and Regions*. Cambridge: Cambridge University Press.

Campbell, J and Pryce, W. 2003. *Brick: a world history*. London: Thames and Hudson.

Cancino, C and Matero, F. 2003. Assessment of Grouting Methods for Cracks and Large-Scale Detachment Repairs, Casa Grande National Monument. *Preprint of Papers of the 9th International Conference on the Study and Conservation of Earthen Architecture, Terra 2003, Yazd, Iran. 29th November – 2nd December 2003*. 395-414. Tehran: Iranian Cultural Heritage Organisation.

Caperton, T. 1987. Fort Selden Ruins Stabilization. In *5th International Meeting of Experts on the Conservation of Earthen Architecture (Rome, 22-23 October 1987)* 13-23. Grenoble: ICCROM/CRATerre-EAG.

Caperton, T. 1990. Fort Selden Ruins Conservation. *6th International Conference on the Conservation of Earthen Architecture, Las Cruces, New Mexico, USA. October 14-19, 1990. Preprints*. 209-211. Los Angeles: Getty Conservation Institute.

Caperton, T. 1993. Long Term Preservation Issues Related to Earthen Archaeological Sites. *7a conferência internacional sobre e estudo e conservação da arquitectura de terra, Silves, Portugal, 24-29 October 1993*. 324-329. Portugal: Direccao Geral dos Edificios e Monumentos Nacionais.

Carrera, E. 1993. Las Cuarenta Casas Archaeological Zone Consolidation and Maintenance at Chiuahua, Mexico. *7a conferência internacional sobre e estudo e conservação da arquitectura de terra, Silves, Portugal, 24-29 October 1993*. 348-352 Portugal: Direccao Geral dos Edificios e Monumentos Nacionais.

Carroll, S. 1998. Temporary production of a tel site excavation in central Turkey. *Conservation and Management of Archaeological Sites*. 2. 155-162.

Carter, T and Pagliero, R. 1966. Notes on Mudbrick Preservation. *Sumer*. 22. 65-76.

Casoli, A. Chiari, G. Burger, R. Salazar-Burger, L. Vizzari, M. and Palla, G. 2000. Identification of the binder in the paintings of the ancient Manchay culture (Peru). *Terra 2000: 8th International Conference on the study and conservation of earthen architecture. Preprints. Torquay, United Kingdom, 11-13 May 2000*. 120-124. London: English Heritage.

Castellanos, C and Hoyle, A. 2000. Conservation management planning for earthen architecture Chan Chan, Peru. *Terra 2000: 8th International Conference on the study and conservation of earthen architecture. Preprints. Torquay, United Kingdom, 11-13 May 2000*. 13-18. London: English Heritage.

Chambers, N. Simmons, C and Wackernagel, M. 2000. Sharing Nature's Interest – ecological footprints as an indicator of sustainability. London: Earthscan.

Chardin, J. *Sir John Chardins Travels in Persia, 1673-1677*. New York: Dover Publications.

Chiari, G. 1980. Treatment of Adobe Friezes in Peru. *Third International Symposium on Mudbrick (adobe) Preservation, 29th September to 4th October 1980, Ankara, Turkey*. 39-45. ICOM/ICOMOS.

Chiari, G. 1987. Consolidation of adobe with ethyl silicate: control of long term effects using sem. *5th International Meeting of Experts on the Conservation of Earthen Architecture (Rome, 22-23 October 1987)*. 25-32. Grenoble: ICCROM/CRATerre-EAG.

Chiari, G. 1990a. Evaluation of the preservation work on earthen architecture done in Iraq in the years 1969-71. *Mesopotamia*. 25. 217-227.

Chiari, G. 1990b. Chemical Surface Treatments and Capping Techniques of Earthen Structures: A Long-Term Evaluation. *6th International Conference on the Conservation of Earthen Architecture, Las Cruces, New Mexico, USA. October 14-19, 1990. Preprints*. 267-273. Los Angeles: Getty Conservation Institute.

Chiari, G, Invernizzi, A and Bertolotto, G. 1993. Investigation and Restoration of Clay Fragmentary Statues from Old Nisa, Turkmenistan. *7a conferência internacional sobre e estudo e conservação da arquitectura de terra, Silves, Portugal, 24-29 October 1993*. 228-230. Portugal: Direccao Geral dos Edificios e Monumentos Nacionais.

Chiari, G, Burger, R and Salazar-Burger, L. 2000. Treatment of an adobe painted freeze in Cardal, Peru and its evaluation after 12 years. *Terra 2000: 8th International Conference on the study and conservation of earthen architecture. Preprints. Torquay, United Kingdom, 11-13 May 2000*. 216-217. London: English Heritage.

Chitty, G. 1987. A Prospect of Ruins, *Transactions of the Association for Studies in the Conservation of Historic Buildings,* 12. 43-60.

Choay, F. 2001. *The Invention of the Historic Monument.* Cambridge: Cambridge University Press.

Clark, K. 1999. *Conservation Plans in Action: Proceedings of the Oxford Conference.* London: English Heritage.

Clark, K. 2001. *Informed Conservation. Understanding historic buildings and their landscapes for conservation.* London: English Heritage.

Clark, K. 2006. From significance to sustainability. In Clark, K (ed). *Capturing the Public Value of Heritage.* 29-30. London: English Heritage.

Clarke, S and Engelbach, R. 1930. *Ancient Egyptian Masonry: The Building Craft.* London: Oxford University Press.

Clayton, J. 2005. Portrait of a village in need of everything G8 can offer.
http://www.timesonline.co.uk/article/0,,22649-1686855,00.html

Cleere, H. 1989. Introduction: the rationale of archaeological heritage management. In H, Cleere (ed). *Archaeological Heritage Management in the Modern World.* One World Archaeology 9, 1-19. London: Routledge.

Cleere, H. 1995. The evaluation of authenticity in the context of the world heritage convention. In *Nara Conference on Authenticity in relation to the world heritage convention. 1-6 November 1994.* 57-66. Paris: UNESCO World Heritage Centre.

Cleere, H. 2000. The World Heritage Convention in the Third World, in McMannon, F and Hatton, A (eds). *Cultural Resource Management in Contemporary Society.* One World Archaeology 33. 99-106. London: Routledge.

Clifford, S and King, A. 1996. *From place to PLACE maps and Parish and Maps.* London: Common Ground.

Coffman, R. Selwitz, C and Agnew, N. 1990. The Getty Adobe Research Project at Fort Selden II. A study of the interaction of Chemical Consolidants with Adobe and Adobe Constituents. *6th International Conference on the Conservation of Earthen Architecture, Adobe 90 Preprints (Las Cruces, New Mexico, USA, October 14-19, 1990).* 250-254. Los Angeles: The Getty Conservation Institute.

Cooke, L. 2002. Mitigating the Impact of Excavations. Unpublished MA Dissertation. Institute of Archaeology, University College London.

Cooke, L. 2003. To Fill or Not to Fill? Retrospective Backfilling at Merv, Turkmenistan. *Preprint of Papers of the 9th International Conference on the Study and Conservation of Earthen Architecture, Terra 2003, Yazd, Iran. 29th November – 2nd December 2003.* 102-109. Tehran: Iranian Cultural Heritage Organisation.

Cooke, L. 2004. Earthen Building Materials and Techniques at Merv, Turkmenistan. In Lehm 2004. 4. *Internationale Fachtagung für Lehmbau, 29.-30 October 2004. Leipzig, Germany.* 52-61. Weimar Germany: Dachverband Lehm e.V.

Cooke, L. Forthcoming. *Merv Backfilling Report.* Ancient Merv Project Technical Report Series.

Coppack, G. 2001. Conservation 'As Found': The Repair and Display of Wigmore Castle, Herefordshire. In McManus, P (ed). *Archaeological Displays and the Public.* 2001. 145-156. London: Archetype Press.

Corfield, M. 1996. Preventive conservation for archaeological sites. In Roy, A and Smith, P (eds). *Archaeological conservation and its consequences. Preprints of the contributions to the Copenhagen Congress, 26-30 August 1996.* 38-42. London: International Institute for Conservation.

Creamer, H. 1990. Aboriginal Perceptions of the Past: The implications for Cultural Resource Management in Australia, in Gathercole, P and Lowenthal, D (eds). *The Politics of the Past*, 130-140. London: Unwin Hyman.

Crosby, A. 1980. Conservation of Painted Lime Plaster on Mud Brick walls at Tumacacori National Monument, USA. *Third International Symposium on Mudbrick (adobe) Preservation, 29th September to 4th October 1980, Ankara, Turkey.* 59-78. ICOM/ICOMOS.

Crinson, M. 1996. *Empire Building Orientalism and Victorian Architecture.* London: Routledge.

Cuneo, P. 1993. The mud-brick architecture of Ghadames in Libya. *7a conferência internacional sobre e estudo e conservação da arquitectura de terra, Silves, Portugal, 24-29 October 1993.* 129-133. Portugal: Direccao Geral dos Edificios e Monumentos Nacionais.

Dachverband Lehm e.V. 2004. *Building with earth – consumer information.* Weimar: Dachverband Lehm e.V.

Damluji, S. 1992. *The valley of mud brick architecture.* London: Garnet.

Davey, N. 1961. *A History of Building Materials.* London: The Camelot Press.

Dekhordi, A. 2004. Abianeh, a village made of earth. In *Lehm 2004. 4. Internationale Fachtagung für Lehmbau,*

29.-30 October 2004. Leipzig, Germany. 28-31.Weimar Germany: Dachverband Lehm e.V.

Demas, M. 1997. Ephesus. In De la Torre, M (ed). *The conservation of archaeological sites in the Mediterranean region. An international conference organised by the Getty Conservation Institute and the J .Paul Getty Museum, 6-12 May 1995.* 127-149. Los Angeles: Getty Conservation Institute.

Demas, M. 2002. Planning for conservation and management of archaeological sites: a values-based approach. In Palumbo, G and Teutonico, J M (eds). *Management planning for archaeological sites: an international workshop organized by the Getty Conservation.* 27-56. Los Angeles: Getty Conservation Institute.

Demas, M. Agnew, N. Waane, S. Podany, J. Bass, A and Kamamba, D. 1996. Preservation of the Laetoli Hominid Trackway in Tanzania. In Roy, A and Smith, P (eds). *Archaeological Conservation and its Consequences. Preprints of the Contributions to the Copenhagen Congress, 26-30 August 1996,* 38-42. London: International Institute for Conservation.

Denslagen, W. 1993. Restoration Theories, East and West, *ASCHB Transactions* 18, 3-7.

Denslagen, W. 1994. *Architectural restoration in* Western *Europe: Controversy and continuity.* Amsterdam: Architectura and Natura Press.

Dethier, J. 1982. *Down to Earth: mud architecture. An Old Idea, a New Future.* London: Thames and Hudson.

Dimacopoulos, J. 1985. Anastylosis and Anasteloseis, *ICOMOS information*, Jan/March 1/1985.

Dowdy, K and Taylor, M. 1993. Investigations into the benefits of site reburial in the preservation of prehistoric plasters in archaeological ruins. *7a conferência internacional sobre e estudo e conservação da arquitectura de terra, Silves, Portugal, 24-29 October 1993.* 480-487. Portugal: Direccao Geral dos Edificios e Monumentos Nacionais.

Dresner, S. 2002. *The Principles of Sustainability.* London: Earthscan.

Dube, B and Ndoro, W. 1993. Preservation of Dhaka structures from archaeological sites. *7a conferência internacional sobre e estudo e conservação da arquitectura de terra, Silves, Portugal, 24-29 October 1993.* 452-453. Portugal: Direccao Geral dos Edificios e Monumentos Nacionais.

Duncan, R. 1947. *Home-made Home.* London: Faber and Faber

Easton, D. 1996. *The Rammed Earth House.* White River Junction, Vermont and Totnes, England: Chelsea Green Publishing Company.

Egeland, P. 1988. *Cob and Thatch.* Exeter: Devon Books.

Emrick, M and Meinhardt, C. 1990. The reconstruction of Traditional Structures in the Al-Turiaf Quarter. *6th International Conference on the Conservation of Earthen Architecture, Las Cruces, New Mexico, USA. October 14-19, 1990. Preprints.* 153-158. Los Angeles: Getty Conservation Institute.

Eshun, E. 2005. G8 special: My Africa. http://www.sundayherald.com/50683

Espinosa, I. 1993. Study of the materials used in the earthen walls of the city of Granada (Spain). *7a conferência internacional sobre e estudo e conservação da arquitectura de terra, Silves, Portugal, 24-29 October 1993.* 464-468. Portugal: Direccao Geral dos Edificios e Monumentos Nacionais.

Faccena, D. 1976. Italian Archaeological Missions of IsMEO. Notes and Conservation Works. *First International Conference on the Conservation of Mud-Brick Monuments. 25th-30th November 1972, Yazd, Iran.* 45-47. Ministry of Culture and Arts, Iran, and UNESCO.

Fathy, H. 1973. *Architecture for the poor, an experiment in rural Egypt.* Chicago and London: The University of Chicago Press.

Ferm, R. 1990. Synthetic Latex-Soil Slurry, A New Adobe Preservation Technique. *6th International Conference on the Conservation of Earthen Architecture, Las Cruces, New Mexico, USA. October 14-19, 1990. Preprints.* 274-276. Los Angeles: Getty Conservation Institute.

Fielden, B. 1994. *Conservation of Historic Buildings.* London: Butterworth-Heineman.

Fielden, B and Jokilehto, J. 1993. *Management Guidelines for World Cultural Heritage Sites.*ICCROM, UNESCO, ICOMOS.

Fiero, K. Matero, F and Rivera, A. 2000. Preservation of prehistoric earthen architectural finishes in Mesa Verde National Park, Colorado. *Terra 2000: 8th International Conference on the study and conservation of earthen architecture. Preprints. Torquay, United Kingdom, 11-13 May 2000.* 31-38. London: English Heritage.

Fodde, E. 2003. An approach to Archaeological Sites in Central Asia: The conservation of Otrar Tobe, Kazakhstan, Selection and Evaluation of Repair Material. *Preprint of Papers of the 9th International Conference on the Study and Conservation of Earthen Architecture, Terra 2003, Yazd, Iran. 29th November – 2nd December 2003.* 170-183. Tehran: Iranian Cultural Heritage Organisation.

Ford, D. Demas, M. Agnew, N. Blanchette, R. Maekawa, S. Taylor, M and Dowdy, K. 2004. Chaco Canyon Reburial program. In *Conservation and Management of Archaeological Sites*, Vol. 6. 177-202.

Frampton, K. Correa, C and Robson, D. 2001. *Modernity and Community: Architecture in the Islamic World.* London: Thames and Hudson.

Frishman, M and Khan, H. 1994. *The mosque: history, architectural development & regional diversity.* London: Thames and Hudson.

Gangler, A. Gaube, H and Petruccioli, A. 2004. *Bukhara, the eastern dome of Islam, urban development, urban space, architecture and population.* Stuttgart and London: Edition Axel Menges.

Glasson, J. Thenvel, R and Chadwick, A. 1994. *Introduction to Environmental Impact Assessment principles and procedures, process, practise and prospects.* London: UCL Press.

Gil, A. 1997. Studies of materials and foundation of Peruvian prehispanic monuments. In Viggiani, C (ed). *Geotechnical Engineering for the Preservation of Monuments and Historic Sites*. 61-68. Rotterdam: Balkema.

Goodman, M. 2002. Site Preservation at Gordion, an Iron Age city in Anatolia. *Conservation and Management of Archaeological Sites,* No 4, Volume 5. 195-212.

Goldsworthy, A. 2000. *Time.* London: Thames and Hudson.

Gosden, C. 1994. *Social Being and Time*. Oxford: Blackwell.

Grabar, O. 1989. Report on Soviet Central Asia. In Serageldin, I (ed). *Space for Freedom*. 283-287. London: Butterworth.

Groom, S. 2004. *A Secret History of Clay: From Gaugin to Gormley.* London: Tate Publishing.

Gupta, D. 2003. Conserving the Heritage in Ladakh case study of Basgo temples. *Preprint of Papers of the 9th International Conference on the Study and Conservation of Earthen Architecture, Terra 2003, Yazd, Iran. 29th November – 2nd December 2003.* 221-254. Tehran: Iranian Cultural Heritage Organisation.

Hall, C and McArthur, S. 1998. *Integrated Heritage Management – principles and practice.* London: HMSO.

Hallet, I and Samizay, R. 1980. *Traditional architecture of Afghanistan.* New York: Garland STPM Press.

Hartzler, R. 1996. *Acrylic Modified Earthen Mortar.* Intermountain Cultural Resource Center Professional paper No. 61. Santa Fe, New Mexico.

Hartzler, R and Oliver, A. 2000. Learning from the site. Evolution of the Fort Union Strategic Preservation Plan. In *Terra 2000: 8th International Conference on the study and conservation of earthen architecture. Preprints. Torquay, United Kingdom, 11-13 May 2000.* 39-45. London: English Heritage.

Hawemann, F. 2004. Renovation of a prefab-housing flat using clay panels to improve the living environment and to increase acceptance. In *Lehm 2004. 4. Internationale Fachtagung für Lehmbau, Leipzig, Germany.* 184-189. Weimar Germany: Dachverband Lehm e.V.

Helmi, F. 1990. Detoriation and Conservation of some Mud Brick in Egypt. *6th International Conference on the Conservation of Earthen Architecture, Las Cruces, New Mexico, USA. October 14-19, 1990. Preprints*. 277-282. Los Angeles: Getty Conservation Institute.

Helwing, B. 2003. Cities of Clay. *Preprint of Papers of the 9th International Conference on the Study and Conservation of Earthen Architecture, Terra 2003, Yazd, Iran. 29th November – 2nd December 2003.* 255-261. Tehran: Iranian Cultural Heritage Organisation.

Herrmann, G. 1999, *Monuments of Merv, Traditional Buildings of the Karakum.* London: Society of Antiquaries.

Herrmann, G. Coffey, H. Laidlow, S and Kurbansakhtov, K. 2002. *The Monuments of Merv, Photographs, Plans and Descriptions of the major monuments of the Merv Oasis.* London: Institute of Archaeology and the British Institute of Persian Studies.

Hillenbrand, R. 1994. *Islamic Architecture*. Edinburgh: Edinburgh University Press.

Hobsbawn, E. 1983. Introduction: Inventing Traditions, in Hobsbawn, E and Ranger, T (eds). *The Invention of Tradition*, 1-14. Cambridge: Cambridge University Press.

Holmes, J. 1975. A Study of Social Organization in Certain Villages in West Khurasan. Iran, with special reference to Kinship and Agricultural activities. Unpublished PhD Thesis, University of Durham.

Horne, L. 1983. Recycling in an Iranian village: Ethnoarchaeology in Baghetsan. *Archaeology* 35 (4). 16-21.

Horne, L. 1993. Occupational and locational instability in arid land settlement. In Cameron, C and Tomka, S (eds). *The Abandonment of Settlements and Regions.* 43-54. Cambridge University Press: Cambridge.

Horne, L. 1994. *Village Spaces, Settlement and Society in Northeastern Iran.* Washington and London: Smithsonian Institution Press.

Houben, H and Guillaud, H. 1994. *Earth Conservation. A comprehensive guide.* London: Intermediate Technology Publications.

Hoyle, A. Carcelen, A and Saavedra, F. 1993. Conservation of the Tomaval Castle. *7th international conference of the study and conservation of earthen architecture. Silves, Portugal, 24-29 October 1993.* 222-227. Portugal: Direccao Geral dos Edificios e Monumentos Nacionais.

Hughes, R. 1983. Techniques and Materials: Material and Structural Behaviour of Soil Constructed Walls. *Monumentum*, 26, No 3. September 1983. 175-188.

Hughes, R. 1988. The geotechnical study of soils used as structural materials in historic monuments. In Maronos, P and Koukis, G (eds). *The engineering geology of ancient work, monuments and historical sites. Preservation and protection, proceedings of an international symposium, Athens, 19-23 Sept. 1988.* 1041-1048. Rotterdam: Balkema.

Hughes, R. 1996. A review of works at the World Heritage Monument of Moenjodaro, Pakistan. In Corfield, M. Hinton, P. Nixon, T and Pollard, M (eds). *Preserving Archaeological Remains in Situ Proceedings of the conference 1st to 3rd April, 1996.* 26-30. London: Museum of London.

Hughes, R. 2002. Method Statement for archaeologically excavating, documenting and analysing buried soil walls. *Terra 2000: 8th International Conference on the study and conservation of earthen architecture. Postprints. Torquay, Devon, UK, May 2000.* 35-43. London: English Heritage.

Hurd, J. 2003. Stories of the Dead, Towards an Understanding of the Traditional Earthen Architecture of Kyrgyzstan. *Preprint of Papers of the 9th International Conference on the Study and Conservation of Earthen Architecture, Terra 2003, Yazd, Iran. 29th November – 2nd December 2003.* 293-295. Tehran: Iranian Cultural Heritage Organisation.

Hurd, J and Gourley, B. 2000. *Terra Britannica: a celebration of earthen structures in Great Britain and Ireland.* London: James and James.

Imankuluv, J and Tentieva, A. 2003. Preservation of Silk Road Sites in Kyrgystan: Navekat, Suyub, Burana. *Preprint of Papers of the 9th International Conference on the Study and Conservation of Earthen Architecture, Terra 2003, Yazd, Iran. 29th November – 2nd December 2003.* 593-598. Tehran: Iranian Cultural Heritage Organisation.

Ito, N. 1995. "Authenticity" Inherent in Cultural heritage in Asia and Japan, in *Nara Conference on Authenticity in relation to the world heritage convention, Nara, Japan, 1-6th November 199,* 35-45. UNESCO World Heritage Centre, Agency for Cultural Affairs, Japan, ICCROM and ICOMOS.

Jaggard, W. 1921. *Experimental Cottages. A report on the work of the Department at Amesbury, Wiltshire.* HMSO: Department of Scientific and Agricultural Research.

Jansen, M. 2003. Mohenjo-Daro, Pakistan, The Problems of Adobe and Brick Conservation. *Preprint of Papers of the 9th International Conference on the Study and Conservation of Earthen Architecture, Terra 2003, Yazd, Iran. 29th November – 2nd December 2003.* 309-318. Tehran: Iranian Cultural Heritage Organisation.

Jerome, P. 1993. Analysis of bronze age mudbricks from Palaikastro, Crete. *7a conferência internacional sobre e estudo e conservação da arquitectura de terra, Silves, Portugal, 24-29 October 1993.* 381-386. Portugal: Direccao Geral dos Edificios e Monumentos Nacionais.

Jerome, P. 2000. The use of lime plasters for waterproofing and decoration of mudbrick buildings in Yemen. *Terra 2000: 8th International Conference on the study and conservation of earthen architecture. Preprints. Torquay, United Kingdom, 11-13 May 2000.* 144-149. London: English Heritage.

Jerome, P. Al-Radi, S and Conlon, J. 2003. Preservation of the Mudbrick Palaces of Tarim, Yemen. *Preprint of Papers of the 9th International Conference on the Study and Conservation of Earthen Architecture, Terra 2003, Yazd, Iran. 29th November – 2nd December 2003.* 319-328. Tehran: Iranian Cultural Heritage Organisation.

Jiyao, H and Weitung, J. 1990. Earth Culture Architecture - The Protection and Development of Rammed Earth and Adobe Architecture in China. *6th International Conference on the Conservation of Earthen Architecture, Las Cruces, New Mexico, USA. October 14-19, 1990. Preprints.* 72-76. Los Angeles: Getty Conservation Institute.

Jokilehto, J. 1999. *A History of Architectural Conservation.* London: Butterworth-Heinemann.

Kahn, L. 1973. *Shelter.* California: Shelter Publications.

Kains-Jackson, C. 1880. *Our Ancient Monuments and the Land Around Them, with a preface by Sir John Lubbock.* London: Elliot Stock.

Kamamba, D. 1990. Cyclical Maintenance of Earthen Architecture as a Future Policy in Tanzania. *6th International Conference on the Conservation of Earthen Architecture, Las Cruces, New Mexico, USA. October 14-19, 1990. Preprints.* 77-80. Los Angeles: Getty Conservation Institute.

Kapfinger, O. and Rauch, M. 2001. *Martin Rauch Rammed Earth*. Berlin: Birkhäuser.

Kemp, B. 2000. Soil (including mud-brick architecture). In Nicholson, P and Shaw, I (eds). *Ancient Egyptian Materials and Technology*. 78-103. Cambridge: Cambridge University Press.

Kenyon, K. 1981. *Excavations at Jericho. Vol.3, The architecture and stratigraphy of the Tell.* London : British School of Archaeology in Jerusalem.

Kerr, J. 1996. *The Conservation Plan: a Guide to the Preparation of Conservation Plans for Places of European Cultural Significance.* New South Wales: National Trust for Australia.

Kezhong, H. Huaiyong, J. Run. C and Lijuan, F. 1990. The Weathering Characteristics of the rocks of the Kezier Grottoes and Research into their Conservation. *6th International Conference on the Conservation of Earthen Architecture, Las Cruces, New Mexico, USA. October 14-19, 1990. Preprints*. 283-288. Los Angeles: Getty Conservation Institute.

Khademzadeh, M. 2003. Change and Continuity in Yazd, Iran. *Preprint of Papers of the 9th International Conference on the Study and Conservation of Earthen Architecture, Terra 2003, Yazd, Iran. 29th November – 2nd December 2003.* 427-439. Tehran: Iranian Cultural Heritage Organisation.

Khademzadeh, M. 2004. Continuity and Change in the City of Yazd. In Jodidio, P (ed). *Iran: Architecture for Changing Societies.* 27-30. Torino: Umberto Allemandi & C.

Khalili, N. 1983. *Racing Alone.* Hesperia: CalEarth Press.

Khalili, N. 1986. *Ceramic Houses: How to build your own.* London: Harper Collins.

Kienzle, P. 1997. Conservation and reconstruction at the Palace of Minos at Knossos. *Conservation and management of Archaeological Sites.* 2. 121-124.

Koob, S. Rogers, M and Sams, G. 1990. Preserving the Eight century B.C Mudbrick architecture at Gordion, Turkey: Approaches to Conservation. *6th International Conference on the Conservation of Earthen Architecture, Las Cruces, New Mexico, USA. October 14-19, 1990. Preprints*. 289-294. Los Angeles: Getty Conservation Institute.

Krist, G. 1937. *Alone through the forbidden land.* Cambridge: Ian Faulkner Publishing.

Langroudi, A. 2003. Polymeric Materials in Restoration & Preservation of Historic Monument. *Preprint of Papers of the 9th International Conference on the Study and Conservation of Earthen Architecture, Terra 2003, Yazd, Iran. 29th November – 2nd December 2003.* 357-362. Tehran: Iranian Cultural Heritage Organisation.

Lewis, H. 1980. Experiments in Mudbrick Conservation at Tepe Nush-I Jan. *Third International Symposium on Mudbrick (adobe) Preservation, 29th September to 4th October 1980, Ankara, Turkey.* 109-118. ICOM/ICOMOS.

Lima, S and Puccioni, S. 1990. General considerations on the preservation of earthen architecture in Minas Gerasi, Brazi; A proposal for re-enforcement of a brayed mud wall structure. *6th International Conference on the Conservation of Earthen Architecture, Las Cruces, New Mexico, USA. October 14-19, 1990. Preprints*. 302-307. Los Angeles: Getty Conservation Institute.

Limbrey, S. 1975. *Soil Science and Archaeology.* London: Academic Press.

Lister, R and Lister, F. 1987. *Aztec Ruins on the Animas, excavated, preserved and interpreted.* Southwest Parks and Monuments Association.

Long, R. 2007. *Walking and Marking.* Edinburgh: National Galleries of Scotland.

Lowenthal, D. 1985. *The Past is a Foreign Country.* Cambridge: Cambridge University Press.

Lui, L. 2004. *The Chinese Neolithic – trajectories to early states.* New York: Cambridge University Press.

Lynas, M. 2004. *High Tide – News from a warming world.* London: Flamingo.

Maas, P. 1993. The architecture of Djenne at the Turn of the Century *7a conferência internacional sobre e estudo e conservação da arquitectura de terra, Silves, Portugal, 24-29 October 1993.* 262-271 Portugal: Direccao Geral dos Edificios e Monumentos Nacionais.

Macaulay, R. 1953. *The Pleasure of Ruins.* New York: Walker.

Macarthy, F. 1994. *William Morris, A life for our time.* London: Faber and Faber.

Maekawa, S. 2004. Monitoring of soil moisture in backfilled soil for conservation of an Ancestral Puebloan great house in Chaco Canyon. In *Conservation and Management of Archaeological Sites*, Vol. 6, no. 3 & 4. 315-324.

Malisius, U. 1993. Ghadames – A historic town in the Northern Sahara. *7a conferência internacional sobre e estudo e conservação da arquitectura de terra, Silves, Portugal, 24-29 October 1993.* 64-67. Portugal: Direccao Geral dos Edificios e Monumentos Nacionais.

Marchand, T. 2000. Walling Old Sanaa, Re-evaluating the resurrection of the city walls. *Terra 2000: 8th*

International Conference on the study and conservation of earthen architecture. Preprints. Torquay, United Kingdom, 11-13 May 2000. 46-51. London: English Heritage.

Marchand, T. 2001. *Minaret Building and Apprenticeship in Yemen.* London: Routledge Curzon.

Markovic, S. 1993. Impermanence/Permanence: Mural Decoration in Northern Ghana, in *7a conferência internacional sobre e estudo e conservação da arquitectura de terra, Silves, Portugal, 24-29 October 1993,*68-72. Portugal: Direccao Geral dos Edificios e Monumentos Nacionais.

Mason, R. 2002. Assessing Values in Conservation Planning: Methodological Issues and Choices. In de la Torre, M (ed). *Assessing the Values of Cultural Heritage.* 5-30. Los Angeles: The Getty Conservation Institute.

Mason, R and Avrami, E. 2002. Heritage values and challenges of conservation planning. In Teutonico, J and Palumbo, G (eds). *Management planning for archaeological sites.* 13-26. Los Angeles: The Getty Conservation Institute.

Masson, V. 1989. Archaeological heritage management in the USSR. In Cleere, H (ed). *Archaeological Heritage Management in the Modern World.* One World Archaeology 9. 196-206. London: Routledge.

Matero, F. 1999. Lessons from the Great House. Condition and treatment history as prologue to site conservation and management at Casa Grande Ruins National Monument. *Conservation and Management of Archaeological Sites 3.* 203-224.

Matero, F. 2000. The conservation of an excavated past. In Hodder, I (ed). *Towards reflexive method in archaeology: the example of Çatalhöyük.* British institute of archaeology at Ankara, monograph no. 28. 71-88. Cambridge: McDonald institute for archaeological research.

Matero, F and Fong, K. 1998. Archaeological site conservation and management an appraisal of recent trends. *Conservation and Management of Archaeological Sites.* 2. 129-142.

Matero, F. del Bono, E. Fong, K. Johansen, R and Barrow, J. 2000. Condition and treatment history as prologue to site conservation at Casa Grande Ruins National Monument, in *Terra 2000: 8th International Conference on the study and conservation of* earthen architecture. *Preprints. Torquay, United Kingdom, 11-13 May 2000*, 52-64. London: English Heritage.

Matero, F and Cancino, C. 2002. The conservation of earthen archaeological heritage. An assessment of recent trends. *Terra 2000: 8th International Conference on the study and conservation of earthen architecture. Postprints. Torquay, Devon, UK, May 2000.* 11-21. London: English Heritage.

Matero, F and Moss, E. 2004. Temporary site protection for earthen walls and murals at Çatalhöyük, Turkey. *Conservation and Management of Archaeological Sites,* Vol 6, no. 3 & 4. 213-227.

Matsikure, J. 2000. The conservation and presentation of earthen structures at Great Zimbabwe. The Posselt house project case study. *Terra 2000: 8th International Conference on the study and conservation of earthen architecture. Preprints. Torquay, United Kingdom, 11-13 May 2000.* 261-265. London: English Heritage.

Matthews, R. 2003. *The Archaeology of Mesopotamia – theories and approaches.* London: Routledge.

Mazar, A. 1999. The conservation and management of mudbrick buildings at Tell Qasile, Israel. *Conservation and Management of Archaeological Sites*, 3. 103-108.

McCann, J. 1983. *Clay and Cob Buildings.* Prince Risborough: Shire.

McCann, J. 1987 Is Clay Lump a traditional building material? *Vernacular Architecture 18.* 1-16.

McHenry, P. 1990. Acoma: A Case Study in Preservation Philosophy and Implementation. *6th International Conference on the Conservation of Earthen Architecture, Las Cruces, New Mexico, USA. October 14-19, 1990. Preprints.* 159-165. Los Angeles: Getty Conservation Institute.

McIntosh, R. 1974. Archaeology and mudwall decay in a West African Village. *World Archaeology.* 6/2. 154-171.

McIntosh, R. 1977. Excavation of mud structures: an experiment from West Africa. *World Archaeology* 9/2. 185-199.

Menon, A. 2003. Conservation as a Critical Activity, in Theophile, E and Gutschow, N (eds). *The Sulima Pagoda, East meets West in the Restoration of a Nepalese Temple,* 106-110. Bangkok: Orchard Press.

Middleton, G. 1951. *Earth Wall Construction III: Adobe (Puddled Earth). Notes on the science of building, 20.* Common wealth experimental building station. Canberra, Australian Government Publishing Service, 1951.

Miele, C. 1998. Real Antiquity and the Ancient Object: The Science of Gothic Architecture and the Restoration of Medieval Buildings. In Brand, V (ed). *The Study of the Past in the Victorian Age,* 103-124. Oxford: Oxbow Books.

Miller, N and Bluemel, K. 1999. Plants and mudbrick: preserving the Midas Tumulus at Gordion, Turkey.

Conservation and Management of Archaeological Sites. 3. 225-237.

Moriset, S. 2001a. Planning of a systematic Monitoring for a selection of monuments at "Ancient Merv" World Heritage Site, Turkmenistan: report on the first mission. Unpublished Report. Grenoble: CRATerre-EAG.

Moriset, S. 2001b. Planning of a systematic Monitoring for a selection of monuments at "Ancient Merv" World Heritage Site, Turkmenistan: report on the second mission. Unpublished Report. Grenoble: CRATerre-EAG.

Murray, P. 2007. Introduction. In Goldsworthy, A. 2007. *Andy Goldsworthy at the Yorkshire Sculpture Park*. 6-11. Breton Hall: Yorkshire Sculpture Park.

Munteanu, S. Amelin, S and Konovalov, V. 2003. Earth Palace Near St. Petersburg. *Preprint of Papers of the 9th International Conference on the Study and Conservation of Earthen Architecture, Terra 2003, Yazd, Iran. 29th November – 2nd December 2003*. 440-445. Tehran: Iranian Cultural Heritage Organisation.

Muoz, G and Bahamondez, M. 1990. Conservacion de un sitio arqueologico construido en tierra. *6th International Conference on the Conservation of Earthen Architecture, Las Cruces, New Mexico, USA. October 14-19, 1990. Preprints*. 371-376. Los Angeles: Getty Conservation Institute.

Nardi, R. 1987a. Consolidation of a mudbrick wall using simple techniques and materials. *ICOM Committee for Conservation. 8th triennial meeting, Sydney, Australia, 6-11 September 198*. 503-507. Los Angeles: Getty Conservation Institute.

Nardi, R. 1987b. Conservation of Medieval Structures of Mudbrick and of Fired Brick in Clay. *5th International Meeting of Experts on the Conservation of Earthen Architecture (Rome, 22-23 October 1987)* 71-80. Grenoble: ICCROM/CRATerre-EAG.

Ndoro, W. 1994. The preservation and presentation of Great Zimbabwe. *Antiquity* 68. 616-623.

Neumann, J and Mehta, P. 1987. Preservation of Adobe Constructions in Rainy Areas. *5th International Meeting of Experts on the Conservation of Earthen Architecture (Rome, 22-23 October 1987)* 103-110. Grenoble: ICCROM/ CRATerre-EAG.

Oates, D. 1990. Innovations in mud-brick: decorative and structural techniques in ancient Mesopotamia. *World Archaeology*. Volume 21, no 3. 388-406.

Oliver, A. 2000. *Fort Selden Adobe Test Wall Project, Phase 1 Final Report*. Los Angeles: Getty Conservation Institute.

Oliver, A and Hartzler, R. 2000. Understanding the deterioration of adobe walls. Fort Union National Monument, New Mexico, USA. *Terra 2000: 8th International Conference on the study and conservation of earthen architecture. Preprints. Torquay, United Kingdom, 11-13 May 2000*. 78-85. London: English Heritage.

Oliver, P. 2003. *Dwellings. The Vernacular House World Wide*. London: Phaidon Press.

Olivier, M. Mesbah, A and Adam, W. 1990. Restauration des murailles de Sana'a, Yemen du Nord Amelioration du Zabour, Methods Traditionnelle de Construction en Terre. *6th International Conference on the Conservation of Earthen Architecture, Las Cruces, New Mexico, USA. October 14-19, 1990. Preprints*. 233-239. Los Angeles: Getty Conservation Institute.

Orazi, R. 2000. Restoration and Maintenance of the earth and stone walls of Khor Rori. *Terra 2000: 8th International Conference on the study and conservation of earthen architecture. Preprints. Torquay, United Kingdom, 11-13 May 2000*. 86-92. London: English Heritage.

Orton, C. Tyers, P and Vince, A. 1993. *Pottery in Archaeology*. Cambridge Manuals in Archaeology. Cambridge: Cambridge University Press.

Othman, Z. 2003. Restoration of al-'Udhaibat and the Royal Palace of Muraba. Preprints. *Moderner Lehmbau 2003*. 81-88. Nachhaltiger Wohnungsbau – Zukunft Okologisches Bauen. Fraunhofer IRB Verlag.

Palumbo, G and Teutonico, J. 2002. *Management planning for archaeological sites: an international workshop organized by the Getty Conservation*. Los Angeles: Getty Conservation Institute.

Patty, R and Minium, L. 1945. *Rammed Earth Walls for Farm Buildings*. Agricultural Engineering Department Agricultural Experiment Station, South Dakota State College. South Dakota State College. Agricultural Engineering Department. Experiment Station bulletin 277. Brookings.

Pearson, G. 1992. *Conservation of Chalk and Clay Buildings*. London: Donhead.

Pearson, D. 2001. *New Organic Architecture*. London: Gaia Books.

Pecararo, A. 1993. The Conservation of the Church of Nossa Senhora do Rosario, Embu, Sao Paulo, Brazil. *7a conferência internacional sobre e estudo e conservação da arquitectura de terra, Silves, Portugal, 24-29 October 1993*. 278-281 Portugal: Direccao Geral dos Edificios e Monumentos Nacionais.

Peek, N. 2004. *Experimentation with different forms of earthen materials conservation, with the aim of designing a conservation strategy for a section through the defensive wall at Merv, Turkmenistan.* Unpublished BSc Dissertation, Institute of Archaeology, University College London.

Piggot, V and Butterbaugh, D. 1978. A Programme in experimental mudbrick preservation at Hasanlu Tepe. *IRAN 16.* 161-168.

Price, C. 2000. Following Fashion: the ethics of archaeological conservation, in F McMannon and A Hatton (eds). *Cultural Resource Management in Contemporary Society,* 213-230. London: Routledge.

Puschnigg, G. 2001. The Pre-Islamic Pottery. In G. Hermann, K. Kurbanskhatov, and St John Simpson *et al.* The International Merv Project Preliminary Report on the Ninth Year (2000). *Iran.* 9-52.

Pwiti, G. 1996. Let the ancestors rest in peace ? New challenges for cultural heritage management in Zimbabawe. *Conservation and Management of Archaeological Sites* 1, 3, 151-160.

Rael, R. 2004. A counter history of modern architecture in the American southwest. In Lehm 2004. 4. *Internationale Fachtagung für Lehmbau, 29.-30 October 2004. Leipzig, Germany.* 66-73.Weimar Germany: Dachverband Lehm e.V.

Rainer, L. 1993. The Tradition of Kassena Wall Painting: Burkina Faso. *7a conferência internacional sobre e estudo e conservação da arquitectura de terra, Silves, Portugal, 24-29 October 1993.* 439-444. Portugal: Direccao Geral dos Edificios e Monumentos Nacionais.

Rapoport, A. 1969. *House Form and Culture.* London: Prentice Hall International.

Rath, R. 2004. Earthen building in the former Soviet occupied zone and the former GDR 1945-1989. In *Lehm 2004. 4. Internationale Fachtagung für Lehmbau 29.-30 October 2004. Leipzig, Germany.* 112-121.Weimar Germany: Dachverband Lehm e.V.

Renfrew, C. 2003. *Figuring it out – the parallel visions of artists and archaeologists.* London: Thames and Hudson.

Reutova, M and Shirinov, T. 2004. Adobe building materials in central Asia. Conservation of archaeological earthen monuments. In Lehm 2004. *4. Internationale Fachtagung für Lehmbau 29.-30 October 2004. Leipzig, Germany.* 170-177.Weimar Germany: Dachverband Lehm e.V.

Richter, A. 2004. An investigation of selected stabilising agents for earthen buildings in arid climates. In Lehm 2004. 4. *Internationale Fachtagung für Lehmbau 29.-30 October 2004. Leipzig, Germany.* 150-161.Weimar Germany: Dachverband Lehm e.V.

Ridley, J. 2003. *Edwin Lutyens His life, his wife, his work.* London: Pimlico.

Riegl, A. 1903. The Modern Cult of the Ruin, In Stanley Price, N, Talley, M and Melucco Vaccaro, A (eds). 1996. *Historical and Philosophical Issues in the Conservation of Cultural Heritage.* 69-83. Los Angeles: The Getty Conservation Institute.

Rivera, A. Culpepper, B. Barrow, J and Fisher, P. 2004. Partial Reburial of West Ruin at Aztec Ruins National Monument. In *Conservation and Management of Archaeological Sites,* Vol. 6. 285-294.

Roca, F. Espinosa, I. Osorio, J. Fernandez, A. Barranco, J. Ortega, E and Martinez, R. 1993. The City Walls of Grenada (Spain) Use, Conservation and Restoration. *7a conferência internacional sobre e estudo e conservação da arquitectura de terra, Silves, Portugal, 24-29 October 1993.* 272-277 Portugal: Direccao Geral dos Edificios e Monumentos Nacionais.

Rojas, E and Crocker, E. 2000. Structuring the discipline - The conflict between theory and practise. *Terra 2000: 8th International Conference on the study and conservation of earthen architecture. Preprints. Torquay, United Kingdom, 11-13 May 2000.* 383-388. London: English Heritage.

Roselund, N. 1990. Repair of Cracked Adobe Walls by Injection of Modified Mud. *6th International Conference on the Conservation of Earthen Architecture, Las Cruces, New Mexico, USA. October 14-19, 1990. Preprints.* 336-341. Los Angeles: Getty Conservation Institute.

Rosen, A. 1986. *Cities of Clay: The Geoarchaeology of Tells. Prehistoric Archaeology and Ecology.* Chicago: University of Chicago Press.

Rothman, H. 1988. *Bandelier National Monument: An Administrative History.* National Park Service Division of History, Southwest Cultural Resources Centre Santa Fe, New Mexico, Professional Papers No. 14. http://www.nps.gov/band/adhi/adhit.htm

Rothschild, N. Mills, B. Ferguson, T and Dublin, S. 1993. Abandonment at Zuni farming villages. In Cameron, C and Tomka, S (eds). *The Abandonment of Settlements and Regions.* 123-137. Cambridge: Cambridge University Press.

Rua, C and Rajer A. 1990. Restoration of the Sixteenth Century Church at Curahuara de Carangas, Bolivia: A case study. *6th International Conference on the Conservation of Earthen Architecture, Las Cruces, New Mexico, USA. October 14-19, 1990 Preprints.* 171-175. Los Angeles: Getty Conservation Institute.

Rua, C. Rajer, A and Mostacedo, N. 1993. A Case study of conservation/restoration of the seventeenth century Adobe Church at Carabuco, Bolivia. *7th international conference of the study and conservation of earthen architecture. Silves, Portugal, 24-29 October 1993*. 205-209. Portugal: Direccao Geral dos Edificios e Monumentos Nacionais.

Rudofsky, B. 1964. *Architecture without architects. A short introduction to non-pedigreed architecture*. New York: Museum of Modern Art.

Ruskin, J. 1849. *The Seven Lamps of Architecture*. (Reprint of the 1880 Edition). New York: Dover Publications.

Said, E. 1993. *Culture and Imperialism*. London: Vintage.

Saunders, A. 1983. A Century of Ancient Monuments Legislation 1882-1982, *The Antiquaries Journal, LXIII*, 11-33.

Scarre, C. 2005. *The Human Past – World Prehistory & the Development of Human Societies*. London: Thames and Hudson.

Schijns, W. 1993. The city of Djenne at the end of the twentieth century. Preservation of the heritage of earthen architecture. *7a conferência internacional sobre e estudo e conservação da arquitectura de terra, Silves, Portugal, 24-29 October 1993*. Direccao 373-378. Portugal: Geral dos Edificios e Monumentos Nacionais.

Schmidt, H. 1997. Reconstruction of Ancient Buildings. In De la Torre, M (ed). *The conservation of archaeological sites in the Mediterranean region. An international conference organised by the Getty Conservation Institute and the J.Paul Getty Museum, 6-12 May 1995*. 41-50. Los Angeles: Getty Conservation Institute.

Schroeder, H. 1993. Traditional Buildings of Lime-Tuff in Thuringia/Germany, in *7a conferência internacional sobre e estudo e conservação da arquitectura de terra, Silves, Portugal, 24-29 October 1993*, 117-122. Portugal: Direccao Geral dos Edificios e Monumentos Nacionais.

Schwartzbaum, P. Silver, C and Wheatley, C. 1980. The Conservation of a Chacolithic Mural Painting on Mud Brick from the site of Teleilat Ghassul, Jordan. *Third International Symposium on Mudbrick (adobe) Preservation, 29th September to 4th October 1980, Ankara, Turkey*. 177-200. ICOM/ICOMOS.

Searle, A. 1912. *The Natural History of Clay*. London: Cambridge University Press.

Seefried, R. 2004. The Heuneburg an der Donau – the earliest use of adobe bricks north of the Alps. In *Lehm 2004. 4. Internationale Fachtagung für Lehmbau 29.-30 October 2004. Leipzig, Germany*. 61-65. Weimar: Dachverband Lehm e.V.

Selwitz, C. 1995. Saving the Fort Selden ruins. The use of a composite blend of chemicals to stabilize fragile historic adobe. *Conservation and Management of Archaeological Sites. 1*. 109-116.

Sengupta, R. 1984. The restoration of the *nirvana-stupa* of the Buddha at Kusinara, India. *Monumentum*. 313-324.

Sharma, R. Gupta, H and Kanotra, Y. 1995. Evaluation of physical characteristics of 'mercula' admixed clay for sealing of cracks in mud plaster of monasteries, Ladkh region. *ICCROM International Colloquium methods of evaluating products for the conservation of porous building materials in monuments. Rome 19th to 21st June 1995*. 41-48. Rome: ICCROM.

Silver, C. 1990. Analyses and Conservation of Pueblo Architectural Finishes in the American SouthWest. *6th International Conference on the Conservation of Earthen Architecture, Las Cruces, New Mexico, USA. October 14-19, 1990. Preprints*. 176-181. Los Angeles: Getty Conservation Institute.

Simpson, W. 1892. Mud, a material in Persian and Eastern Architecture. *Journal of the Society of Arts*. XL 696-714.

Sivan, R. 1997. The Presentation of Archaeological Sites. In De la Torre, M (ed). *The conservation of archaeological sites in the Mediterranean region. An international conference organised by the Getty Conservation Institute and the J.Paul Getty Museum, 6-12 May 1995*. 51-59. Los Angeles: Getty Conservation Institute.

Skibinski, S. 1990. Problemas de la investigacion y conservacion de las estructuras del centro ceremonial de la cultura paracas-naza, Peru. *6th International Conference on the Conservation of Earthen Architecture, Las Cruces, New Mexico, USA. October 14-19, 1990. Preprints*. 443-448. Los Angeles: Getty Conservation Institute.

Solli, B. 2000. Protection of the environment and the role of archaeology. In McManamon, F and Hatton, A (eds). *Cultural Resource Management in Contemporary Society*, 93-98. London: Routledge.

Soliani, C. Almeida, A. Hofer, M and Bitencourt, E. 1993. Taipa de Sopapo na Imigracao Alema no Rio Grande do Sul, Estudo de Caso. *7a conferência internacional sobre e estudo e conservação da arquitectura de terra, Silves, Portugal, 24-29 October 1993*. 454-459. Portugal: Direccao Geral dos Edificios e Monumentos Nacionais.

Sramek, J and Losos, L. 1990. Outline of Mudbrick Structures Conservation at Abusir, Egypt. *6th International Conference on the Conservation of Earthen Architecture, Las Cruces, New Mexico, USA. October 14-19, 1990. Preprints*. 449-454. Los Angeles: Getty Conservation Institute.

Stanley Price, N. Talley, M and Melucco Vaccaro, A. 1996. *Historical and Philosophical Issues in the Conservation of Cultural Heritage.* Los Angeles: The Getty Conservation Institute.

Steele, J. 1997. *An Architecture for People. The Complete works of Hassan Fathy.* London: Thames and Hudson.

Steingass, P. 2003. New Chances for Modern Earth Building, *Moderner Lehmbau 2003.* 186-190 .Nachhaltiger Wohnungsbau – Zukunft Okologisches Bauen. Fraunhofer IRB Verlag.

Stevanovic, S. 1999. Report on experimental archaeology at Çatalhöyük. Çatalhöyük 1999 Archive Report. http://catal.cam.ac.uk

Stevens, A. 1983. Architecture de terre: Monuments et sites de l'oasis de Turfan (Xinjiang) sur la route de la soie. *Monumentum.* 45-69.

Stille, A. 2002. *The Future of the Past: How the Information Age Threatens to Destroy our Cultural Heritage.* London: Picador.

Stohr, K. 2006. 100 Years of Humanitarian Design. In Architecture for Humanity (eds). *Design Like You Give Damn.* 33-55. London: Thames and Hudson.

Sullivan, S. 1993. Conservation Policy Delivery, in MacLean, M (ed). *Cultural Heritage in Asia and the Pacific: Conservation and Policy.* 15-26. Los Angeles: The Getty Conservation Institute.

Sullivan, S. 1997. A planning model for the management of archaeological sites. In De la Torre, M (ed). *The conservation of archaeological sites in the Mediterranean region. An international conference organised by the Getty Conservation Institute and the J.Paul Getty Museum, 6-12 May 1995,* 15-26. Los Angeles: The Getty Conservation Institute.

Swenarton, M. 2003. Rammed earth revival: technological innovation and government policy in Britain, 1905-1925. *Construction History* Vol. 19. 2003. 107-126.

Taylor, M. 1987. Fort Selden Test Wall Status Report. In *5th International Meeting of Experts on the Conservation of Earthen Architecture (Rome, 22-23 October 1987)* 91-102. Grenoble: ICCROM/CRATerre-EAG.

Taylor, M. 1990. An Evaluation of the New Mexico State Monuments Adobe Test Walls at Fort Selden. *6th International Conference on the Conservation of Earthen Architecture, Las Cruces, New Mexico, USA. October 14-19, 1990. Preprints.* 383-389. Los Angeles: Getty Conservation Institute.

Theophile, E and Gutschow, N. 2003. *The Sulima Pagoda, East meets West in the Restoration of a Nepalese Temple.* Bangkok: Orchard Press.

Thompson, M. 1981. *Ruins: their preservation and display.* London: British Museum.

Thomas, G. 1999. The role of research and education in site management at the Lemba Experimental Village, Cyprus. *Conservation and Management of Archaeological Sites* 3. 109-123.

Thönnes, M. 2004. Wood and earth: a healthy combination for contemporary living. In Lehm 2004. *4. Internationale Fachtagung für Lehmbau 29.-30 October 2004. Leipzig, Germany.* 178-183.Weimar Germany: Dachverband Lehm e.V.

Torraca, G. 1980. Introduction. *Third International Symposium on Mudbrick (adobe) Preservation, 29th September to 4th October 1980, Ankara, Turkey.* ICOM/ICOMOS.

Torraca, G. 1981. *Porous Building Materials. Materials Science for Architectural Conservation.* Rome: ICCROM.

Torraca, G. Chiari, G and Gullini, G. 1972. Report on mud-brick preservation, *Mesopotamia,* VII, 259-287.

Trotter, D. 2000. *Cooking with Mud, the idea of mess in nineteenth-century art and fiction.* Oxford: Oxford University Press.

Truscott, M and Young, D. 2000. Revising the Burra Charter: Australia ICOMOS updates its guidelines for conservation practice, *Conservation and Management of Archaeological Sites,* 4, 101-116.

Vatandoust, R. 2003. Introduction. *Preprint of Papers of the 9th International Conference on the Study and Conservation of Earthen Architecture, Terra 2003, Yazd, Iran. 29th November – 2nd December 2003.* Tehran: Iranian Cultural Heritage Organisation.

Vatandoust, R and Mohktari, E. 2004. The earthquake catastrophe in Bam, Iran – future strategies for a destroyed city made of earth. In *Lehm 2004. 4. Internationale Fachtagung für Lehmbau 29-30 October 2004. Leipzig, Germany.* 220-233. Weimar Germany: Dachverband Lehm e.V.

Vitruvius. *The Ten Books on Architecture.* (1960 Edition, translated by M Morgan, M). New York: Dover Publications.

Walker, B and McGregor, C. 1996a. *The Hebridean Blackhouse a guide to materials, construction and maintenance. Historic Scotland Technical Advice Note 5.* Edinburgh: Historic Scotland.

Walker, B and McGregor, C. 1996b. *Earth Structures and Construction in Scotland. Historic Scotland Technical Advice Note 6.* Edinburgh: Historic Scotland.

Walker, P. Keable, R. Marton, J and Maniatidis, V. 2005. *Rammed Earth: Design and Construction Guidelines.* Watford: Building Research Establishment.

Walls, A. 2004. Arabian Mud Brick Technology: Some Thoughts After the Bam Earthquake. *Construction History Society Newsletter.* 69 July 2004. 11-20

Warren, J. 1993. *Earthen Architecture.* ICOMOS.

Warren, J. 1999. *Conservation of Earth Structures.* London: Butterworth-Heinemann.

Weiamann, A and Bryce, K. 2006. *Building with Cob.* Foxhole, Devon: Green Books.

Wetherall, D. 1998. The Growth of Archaeological Societies. In Brand, V (ed). *The Study of the Past in the Victorian Age*, 21-34. Oxford: Oxbow Books.

Wheeler, G. 2005. *Alkoxysilanes and the consolidation of stone.* Los Angeles: Getty Conservation Institute.

Williams-Ellis, C. 1920 *Cottage Building in Cob, Pisé, Chalk, and Clay, a renaissance* London: Country Life.

Williams-Ellis, C. 1946. *The Adventure of Building.* London: The Architectural Press.

Wilson, C. 1997. *The Myth of Santa Fe, creating a modern regional tradition.* Albuquerque: University of New Mexico Press.

World Health Organisation. 2003. *Climate change and human heath - risks and responses.* WHO.

Wulff, H. 1966. *The traditional crafts of Persia, their development, technology and influence in Eastern and* Western *civilizations*. Cambridge, Massachusetts: Cambridge Massachusetts Institute of Technology.

Zuixiong, L. 1990. Consolidation of a Neolithic Earthen Site with Potassium Silicate. *6th International Conference on the Conservation of Earthen Architecture, Las Cruces, New Mexico, USA. October 14-19, 1990. Preprints*. 295-301. Los Angeles: Getty Conservation Institute.

Other Resources

Can technology make poverty history? Report online accessed 8/8/2007. http://www.channel4.com/science/microsites/M/makepovertyhistory/barefoot/

DRUK School Website http://www.dwls.org/

Oxford English Dictionary (OED). accessed 8/8/2007 Through North Yorkshire County Council Library Access http://dictionary.oed.com

Before and after photographs of the Arg-e Bam searchable at: http://news.bbc.co.uk/1/hi/magazine/in_pictures/3422997.stm.

Charter and Recommendation (chronological)

The Manifesto for the Society for the Protection of Ancient Buildings. 1877.

Recommendations of the Madrid Conference (Sixth International Congress of Architects) (1904) *The Architectural Journal Being the Journal of the Royal Institute of British Architects (RIBA)* Vol. XI. Third Series, 1904. 343-346.

The Athens Charter for the Restoration of Historic Monuments. 1931.

UNESCO Recommendation on International Principles Applicable to Archaeological Excavation. 1956.

The Venice Charter: International Charter for the Conservation and Restoration of Monuments and Sites. 1964.

UNESCO Convention Concerning the Protection of the World Cultural and Natural Heritage. 1972.

The Burra Charter: The Australia ICOMOS Charter for the Conservation of Places of Cultural Significance, 1981. (Revised 1999)

World Commission on Environment and Development, 1987. *Our Common Future.* Oxford: Oxford University Press. (= Brundtland Report)

The Vermillion Accord on Archaeological Ethics and the Treatment of the Dead. 1989.

ICOMOS Charter for the Protection and Management of the Archaeological Heritage. 1990.

Native American Grave Protection and Repatriation Act (NAGPRA), 1990.

ICOMOS New Zealand Charter for the Conservation of Places of Cultural Heritage Value. 1992.

Agenda 21: The United Nations Project for Action from Rio. United Nations Dept of Information. New York. 1993.

Nara Document on Authenticity. 1994. *Nara Conference on Authenticity in relation to the world heritage convention. Proceedings Nara Conference on Authenticity in relation to the world heritage convention, Nara, Japan, 1-6 November 1994.* Paris: UNESCO World Heritage Centre.

The San Antonio Declaration, ICOMOS. 1996.

English Heritage. *Reconstructions on Archaeological Sites.* Draft Guidelines. 1999.

UNESCO Convention for the Safeguarding of the Intangible cultural heritage. 2003.

UNESCO World Heritage Indigenous Peoples Council of Experts (WHIPCOE) http://whc.unesco.org/whipcoe/

INTACH (Indian National Trust for Art and Cultural Heritage) Charter for the Conservation of Unprotected Architectural Heritage and Sites in India. 2004.

International Conferences Concerned with the Study and Conservation of Earthen Architecture.

1st International Conference on the Conservation of Mud-Brick Monuments. 25th -30th November 1972, Yazd, Iran. ICOMOS.

2nd International Conference on the Conservation of Mud-Brick Monuments. 1976, Yazd, Iran. ICOMOS.

3rd International Symposium on Mudbrick (adobe) Preservation. 29th September to 4th October 1980, Ankara, Turkey.

4th International conference on Mudbrick Preservation, Lima, Peru. 1984.

5th International Meeting of Experts on the Conservation of Earthen Architecture. 22nd-23rd October 1987, Rome, ICCROM/CRATerre-EAG, Grenoble.

6th International Conference on the Conservation of Earthen Architecture. October 14th –19th 1990, Las Cruces, New Mexico. Getty Conservation Institute. Los Angeles.

7th International conference of the study and conservation of earthen architecture. 24th –29th October 1993, Silves, Portugal.

8th international conference on the study and conservation of earthen architecture. 11th –13th May 2000, Torquay, United Kingdom. English Heritage. London.

9th International Conference on the Study and Conservation of Earthen Architecture, 29th November – 2nd December 2003, Yazd, Iran. Iranian Cultural Heritage Organisation, Tehran.

INDEX OF MAIN SITES

www.ingramcontent.com/pod-product-compliance
Lightning Source LLC
LaVergne TN
LVHW072255300926
849160LV00018B/1672
* 9 7 8 1 4 0 7 3 0 6 8 8 9 *